Bridging the Digital Divide

NIPA® GENX ELECTRONIC RESOURCES & SOLUTIONS P. LTD.
New Delhi-110 034

About the Editors

Dr. Vishnu D. Rajput, a renowned Principal Scientist, currently head, International Laboratory of Nanobiotechnology at the Academy of Biology and Biotechnology, Southern Federal University, Rostov-on-Don, Russia. With a wealth of experience and experimental work, Dr. Rajput has made significant strides in environmental science research, particularly in ensuring food safety in a sustainable manner. He has published 376 scientific publications, 88 chapters, 28 books, and H-Index: 40. He is holding national and international projects including Mega- Grant, BRICS & Priority 2030. Dr. Rajput has received "Certificate for Appreciation 2019" 2021, "Certificate of Honor 2020", Diploma Award 2021 & 2022, Letter of Gratitude 2022 and 2023 by Southern Federal University, Russia, for outstanding contribution in academic, creative research, and publication activities. He has also received "Highly Qualified Specialist status by Russian government. In year 2023, he is included in the list of 2% worlds potential scientists by Elsevier BV publishing house.

Dr. R.K. Doharey, a distinguished academician and researcher, holds a Master of Science (Agriculture), NET, Ph.D., FSSDAT, FSRDA, and FISEE. He is currently the Professor and Head of the Department of Extension Education at Acharya Narendra Deva University of Agriculture & Technology, Kumarganj, Ayodhya, Uttar Pradesh, India. Concurrently, he serves as a member of the Board of Management at Mahatma Gandhi University of Horticulture & Forestry, Durg, Chhattisgarh, India

With a professional journey spanning over 24 years, Dr. Doharey has made significant contributions to the field of agriculture and extension education. His academic tenure includes a notable 13-year period as a Professor. His scholarly contributions encompass 140 research papers, 16 authored books, and 15 book chapters, reflecting his extensive knowledge and dedication to his field. His excellence has been recognized with 35 awards, further underscoring his significant influence in the field. Dr. Doharey's international engagements extend to 18 countries, amplifying his reach and influence. His previous roles include Assistant Dean at the College of Agriculture, Azamgarh Campus (ANDUAT, Ayodhya, U.P., India), Subject Matter Specialist in Agricultural Extension at C.S.A. University of Agriculture & Technology, Kanpur, U.P., India, and Science Writer at Uttar Pradesh Council of Agricultural Research, Lucknow, U.P., India.

Dr. Veenita Kumari is currently Deputy Director, Centre for Gender in Agriculture, Nutritional Security & Urban Agriculture at National Institute of Agricultural Extension Management (MANAGE). She is also the Principal Coordinator of Post Graduate Diploma in Agricultural Extension Management (PGDAEM). Prior to joining MANAGE, she had worked as Assistant Professor (Extension & Communication Management) in College of Home Science, Central Agricultural University, Tura, Meghalaya for more than 13 years. At CAU Tura, she was primarily involved in teaching (B.Sc. and M.Sc.), research and outreach activities of the college (on and off-campus). A university gold medalist during her M.Sc., Dr. Veenita has 35 published research articles in refereed academic national and international journals, editor of five books, and three book chapters. Her areas of specialization include Women Empowerment, Gender Studies, Women in Agriculture, Women Entrepreneurship, Community Nutrition, Rural Development, Foods & Nutritional Security, Nutrition Sensitive Agriculture, Biofortification, One Health, Family Farmer, Urban Agriculture and Microgreens. She has received Young Scientist Award and Young Teacher Award for her academic excellence.

Mr. Aman Verma, born in Hardoi, Uttar Pradesh, India. He is an esteemed scholar in the field of Extension Education, holds a B.Sc. (Hons.) and M.Sc. in Agricultural Extension from Kulbhaskar Ashram P.G. College, Prayagraj, India, a NAAC 'A' Grade institution with UGC potential for excellence. He is currently pursuing his Ph.D. in Extension Education at Acharya Narendra Deva University of Agriculture and Technology, Ayodhya, Uttar Pradesh, India.

Mr. Verma has made significant strides in his field. His academic achievements are complemented by his extensive career as an author and researcher. He has qualified for the UGC-NET and has published 3 books, 2 as an associate editor and 1 as a textbook. In addition to his books, he has contributed to 10 book chapters, numerous research papers, review papers, and over 40 articles in popular magazines. His expertise extends beyond writing he has also delivered radio talks and guest lectures. He actively contributes to his field through participation in training programs, workshops, national and international seminars, and conferences. Mr. Verma's outstanding contributions have been recognized with prestigious awards from various societies. His commitment to leveraging technology to empower rural women continues to drive his research and impact his field.

Amrit Warshini, born in Patna, Bihar, is a Ph.D. Scholar in the Department of Extension Education, Acharya Narendra Deva University of Agriculture and Technology, Ayodhya, U.P., India. She completed her post graduation from Jawahar Nehru Krishi Vishwavidyalaya, Jabalpur, Madhya Pradesh, India. She qualified for the ICAR SRF-2023 and ICAR NET-2023 with distinction, her NET qualification and prestigious ICSSR scholarship demonstrates a commitment to academic excellence, while her advocacy for gender and social awareness infuses with her purpose. She has published many book chapters, articles, participated in seminars, trainings, National and International Conferences. She has also delivered radio talks and guest lectures. She truly believes in sharing her knowledge and experiences gained to help and pave way for others.

Bridging the Digital Divide

Empowering Women in Rural India through Emerging Technology

Editors

Vishnu D. Rajput

R.K. Doharey

Veenita Kumari

Aman Verma

Amrit Warshini

NIPA® GENX ELECTRONIC RESOURCES & SOLUTIONS P. LTD.

New Delhi-110 034

NIPA® GENX ELECTRONIC RESOURCES & SOLUTIONS P. LTD.

101,103, Vikas Surya Plaza, CU Block
L.S.C.Market, Pitam Pura, New Delhi-110 034
Ph : +91 11 27341616, 27341717, 27341718
E-mail: newindiapublishingagency@gmail.com
www: www.nipabooks.com

For customer assistance, please contact
Phone: + 91-11-27 34 17 17
Fax: + 91-11-27 34 16 16

Print ISBN: 978-93-58877-60-1
ebook ISBN: 978-93-58877-51-9

Composed and Designed by NIPA®.

डॉ. हिमांशु पाठक
DR. HIMANSHU PATHAK
सचिव (डेयर) एवं महानिदेशक (आईसीएआर)
Secretary (DARE) &
Director General (ICAR)

भारत सरकार
कृषि अनुसंधान और शिक्षा विभाग एवं
भारतीय कृषि अनुसंधान परिषद
कृषि एवं किसान कल्याण मंत्रालय, कृषि भवन, नई दिल्ली–110 001

GOVERNMENT OF INDIA
DEPARTMENT OF AGRICULTURAL RESEARCH AND EDUCATION (DARE)
AND
INDIAN COUNCIL OF AGRICULTURAL RESEARCH (ICAR)
MINISTRY OF AGRICULTURE AND FARMERS WELFARE
Krishi Bhavan, New Delhi 110 001
Tel: 23382629 / 23386711 Fax: 91-11-23384773
E-mail: dg.icar@nic.in

Foreword

The digital revolution sweeping across India presents a unique opportunity to transform the lives of rural women engaged in agriculture. The power of emerging technologies is being used to empower these women, enhance their skills, and drive sustainable development in rural communities. The empowerment of women in agriculture aims to create an enabling environment where rural women can thrive as entrepreneurs, decision-makers, and leaders. Its nuanced approach to leveraging technology as a catalyst for change sets this work apart. The authors delve into innovative solutions like precision agriculture, e-commerce platforms, and digital training modules, highlighting their potential to enhance productivity, improve market linkages, and build essential skills.

This book, "Bridging the Digital Divide: Empowering Women in Rural India through Emerging Technology," comprehensively explores how digital tools can bridge existing gaps and unlock new avenues for progress. The authors have meticulously analysed the socio-economic challenges faced by rural women, including limited access to education, healthcare, and economic opportunities. I hope the book will provide valuable insights and actionable strategies for policymakers, development agencies, researchers, and stakeholders in rural development and women's empowerment initiatives across India. By implementing the insightful recommendations presented here, we can collectively unlock the immense potential of India's rural women.

I commend the authors for their dedication to this critical cause and their rigorous approach in exploring the intersection of technology and women's empowerment. Their work serves as a clarion call to collectively bridge the digital divide, foster inclusive growth, and empower our rural women to become architects of a thriving and inclusive agricultural future for India.

(Himanshu Pathak)

Dated the 8th May, 2024

Dr. P. Chandra Shekara
(Director General, MANAGE)

National Institute of Agricultural Extension Management
(MANAGE)
(An Organization of Ministry of Agriculture & Farmers Welfare, Government of India)
Rajendranagar, Hyderabad-500030, Telangana, India
Contact: 040-24594505, 040-24015253, 040-24015388
Email: dgmanage@manage.gov.in

Foreword

As we stride forward into an increasingly digital world, it is imperative that we ensure no one is left behind. In rural India, where the divide between access to technology and its benefits is often stark, empowering women becomes not just a matter of equality, but a catalyst for sustainable development.

"Bridging the Digital Divide: Empowering Women in Rural India through Emerging Technology" is a timely and indispensable contribution to the discourse on leveraging technology for the upliftment of rural women. This book, meticulously curated and expertly crafted, delves into the multifaceted dimensions of empowering women in rural India through emerging technologies.

MANAGE recognises the significance of such technology-driven interventions as an organization dedicated to strengthening agricultural extension systems. Our efforts align with the vision of creating an enabling environment where rural women can thrive as entrepreneurs, decision-makers, and leaders in the agricultural sector.

This book encapsulates a broad spectrum of topics ranging from digital literacy and skill development to the pivotal roles of women in environmental sustainability. It navigates through the nuances of financial inclusion, telehealth, e-learning, e-commerce, and entrepreneurship, presenting a comprehensive roadmap towards bridging the digital gap.

One of the most commendable aspects of this volume is its focus on the voices and experiences of rural women themselves. By centering their stories, aspirations, and challenges, it ensures that the discourse remains grounded in the realities of those it seeks to empower.

In a world where technology often reflects the biases of its creators, "Bridging the Digital Divide" goes a step further by advocating for the active involvement of women in the design and implementation of technology, particularly in the agricultural sector. By cultivating women's leadership in technology, the book not only promotes gender equity but also enhances the effectiveness and inclusivity of technological solutions.

As the Director General of MANAGE, I am proud to endorse this seminal work, which not only identifies the challenges but also illuminates the vast opportunities that lie ahead. It is my fervent hope that the insights contained within these pages serve as a guiding light for policymakers, practitioners, and stakeholders alike, as we collectively endeavour to empower rural women and build a more equitable and prosperous future for all.

(**Dr. P.Chandra Shekara**)
(Director General, MANAGE)

MAHATMA GANDHI UNIVERSITY OF HORTICULTURE & FORESTRY
SANKRA - PATAN, DURG
Camp Office : Krishak Sabhagar, RAIPUR-492 012 (C.G.) INDIA
mguvv.ac.in vcmguvv@gmail.com, (O) 0771-2991101

No./VC-Cell/ /2024/ 528 Dated 08/ 05 / 2024

Dr. R.S. Kureel
Vice Chancellor

Foreword

Empowering women, particularly in rural India, is a cornerstone of sustainable development and a catalyst for transformative change. As we embrace the fourth industrial revolution, we must leverage the power ot emerging technologies to bridge the digital divide and unlock new avenues and opportunities for women empowerment, improving livelihood of their family, sustainable development of rural communities by enhancing their access to digital literacy.

This insightful book, "Bridging the Digital Divide: Empowering Women in Rural India through Emerging Technology," offers a comprehensive and thought-provoking exploration of how cutting-edge technologies can be harnessed to empower rural women in India. By addressing the multifaceted challenges they face in accessing education, healthcare. and economic opportunities, this book presents a compelling case for integrating technology into rural development initiatives.

The authors have skillfully demonstrated how emerging technologies, such as digital literacy programs, mobile healthcare applications, and e-commerce platforms, can serve as powerful tools in improving the livelihoods of rural women, enhancing their skills and driving sustainable growth within their communities. The insights and recommendations presented in this book are not only thought-provoking but also actionable. providing a roadmap for policymakers, organizations, and stakeholders to leverage technologies are effective for the empowerment of rural women.

As a nation committed to inclusive growth and sustainable development, we are responsible for ensuring that no one is left behind, especially women being most vulnerable and marginalized segments of our society. This book serves as a clarion call to embrace technological advancements as catalysts for positive change, empowering women to become path maker of their own development and contributing to the nation building.

I wholeheartedly recommend this book to policymakers, development professionals, researchers and other stake holders in the intersections of technology, gender equality and rural development. By fostering a deeper understanding of the potential of emerging technologies, we can collectively work towards creating a more inclusive and equitable society, where women are empowered to realize their full potential and contribute to building a more prosperous India.

Place: Raipur
Dated: 8th May 2024

(Dr. R.S. Kureel)
Vice-Chancellor
Mahatma Gandhi University of Horticulture and Forestry Drug, Chhattisgarh, India

Formerly :

- Vice Chancellor, Birsa Agricultural University Kanke, Ranchi, Jharkhand
- Vice Chancellor, Dr. B.R. Ambedkar University of social Sciences, Mhow, Indore M.P.
- Vice Chancellor, Acharya N.D. University of Agriculture & Technology, Ayodhya U.P.
- Member, National Education Policy-2020, Ministry of Education, Govt. of India, New Delhi.

Dr. Sugandha Munshi
Senior Associate Scientist and Lead Specialist
Sustainable Impact Department
International Rice Research Institute
s.munshi@irri.org

Foreword

When it comes to "Women-led developments" in India, rural women constitute the backbone of these endeavors. Rural women are the key drivers of change. They are the nation builders! Working from dawn to dusk they play a crucial role in transforming the social, economic, political, environmental, and developmental ecosystem at the grassroots level. Agriculture is the main sector of the rural Indian economy and the significant contribution made by women in this sector further positions them as a catalyst for the nation's economic growth. Because of the significant contributions women make to agriculture, they will be vital to feeding the 9.3 billion people on the planet by 2050.

There is a "feminization" of the agriculture sector, where women are intensively involved in different roles as farmers, cultivators, entrepreneurs, etc. Rural economy and food security benefit greatly from their substantial contribution. However, there are still barriers to women being recognized in agriculture. In terms of equitable access to productive resources, they fall behind. Hence it becomes crucial to create a conducive ecosystem for her to thrive, rise, and shine.

Nowadays, as the digital revolution permeates every aspect of our lives, there is a growing opportunity for digital platforms and technologies to empower women in agriculture, health, education, and financial inclusion. It holds great promise for giving women access to appropriate information about climate-resilient technologies, agro-advisories, direct market linkage, and network formation. Where rural women's empowerment and development are the main priorities, these inherent potentials must be discovered.

Effectively and systematically, the book "Bridging the Digital Divide: Empowering Women in Rural India through Emerging Technology" presents methods and evidence regarding how digital tools and technologies can be gender-responsive and contribute to the empowerment of women in rural India. Unlocking the enormous potential of women in rural India, the authors have offered perceptive strategies and suggestions for fostering a more equal, fair, and inclusive digital environment. I hope that the book will serve as a torchbearer, offering priceless insights into approaches to the decision-makers, developers, researchers, students, and other stakeholders involved in the empowerment of women in India.

I applaud the book's editors and authors for their tenacity and meticulous work in examining the relationship between women's empowerment and technology.

Dated the 9th of May 2024

Sugandha Munshi

(Sugandha Munshi)

Patna

Prof. (Dr.) Rama
Principal, Hansraj College
University of Delhi, Delhi - 110 007

Foreword

At Hansraj College, we are guided by the vision of pioneers like Mahatma Hansraj and Maharshi Dayanand Saraswati, who emphasized the transformative power of knowledge and education. Keeping with this legacy, we strive to provide a holistic and inclusive learning environment that empowers our students, especially young women, to unlock their full potential and become agents of positive change.

With this ethos, I introduce "Bridging the Digital Divide: Empowering Women in Rural India through Emerging Technology" – a timely and important work that explores how digital innovations can enhance access to education, skills development, and livelihood opportunities for rural women in India.

As an institution of higher education, we understand education's pivotal role in addressing societal challenges and fostering sustainable development. Rural women in India continue to face numerous barriers, including limited access to formal education, healthcare, and economic opportunities. This book presents a compelling case for utilizing emerging technologies to bridge these gaps and unlock rural women's immense potential.

The authors have meticulously examined various digital tools and platforms that can facilitate distance learning, skill-building programs, and entrepreneurial ventures for rural women. By leveraging technologies such as e-learning modules, online vocational training, and e-commerce platforms, we can equip these women with the knowledge, skills, and resources they need to overcome challenges and become catalysts of change in their communities.

This book resonates strongly with Hansraj College's commitment to providing cutting-edge infrastructure and fostering an environment that nurtures academic excellence and extracurricular pursuits. The insights and strategies presented in this work will be invaluable for educational institutions, policymakers, development organizations, and stakeholders working towards women's empowerment and rural development in India.

I commend the authors for their dedication to this crucial issue and visionary approach to harnessing technology for social transformation. This book is a must-read for anyone interested in leveraging the power of education and digital innovations to create a more equitable and prosperous society, where rural women can thrive as leaders, decision-makers, and catalysts of positive change.

Date- 8th May 2024
New Delhi

Prof. (Dr.) Rama

Preface

With the emerging digital approaches, it could be imagine where a woman farmer in a remote village can access real-time market prices on her mobile phone, or a young girl can pursue online education after finishing her household chores. This is the transformative potential of bridging the digital divide for rural women in India, a challenge tackled head-on in this book, "Bridging the Digital Divide: Empowering Women in Rural India through Emerging Digital Technologies." This new book delves into these intricacies, examining the unique challenges faced by rural Indian women and the transformative power of emerging technologies in addressing issues like socioeconomic, cultural, and infrastructural barriers.

Spanning a diverse range of topics (explored in detail in chapters like "Mobile Revolution: Connecting the Last Mile" and "Financial Inclusion: Fintech and Access to Banking Services"), this new book explores the current landscape of women's empowerment in rural India, delving into the roles, aspirations, and traditional constraints that shape their lived experiences.

Recognizing the importance of real-world impact, this new book presents a case study and showcases that 'how technology is uplifting rural women (explored further in Chapter 14: "Case Studies in Women Empowerment"). It fosters a holistic perspective by exploring the intersections of gender, technology, and sustainable development (addressed in Chapter 15: "Role of Women in Environmental Sustainability").

Ultimately, "Bridging the Digital Divide" serves as a call to action for policymakers, technologists, entrepreneurs, and changemakers to join forces in creating a gender-just digital future. By fostering collaboration, knowledge-sharing, and sustainable ecosystems (explored in Chapter 16: "Building Sustainable Ecosystems: Fostering Collaboration and Knowledge Sharing"), this new book unlock the transformative potential of technology and empower rural women to become agents of positive change in their communities and beyond.

This new book is a testament to the resilience, strength, and untapped potential of rural women in India. It celebrates their aspirations and provides a roadmap for harnessing the power of emerging technologies to uplift their lives, foster economic empowerment, and contribute to the nation's overall development. This new book will be valuable asset to bridge the digital divide and pave the way for a more equitable and inclusive digital future.

Regards

Editors

Contents

1

Current Perspective on Empowering Women in Rural India

Pradeep Kumar Yadav and Nalin Kumar Mishra

Department of Agricultural Extension, Tilak Dhari Postgraduate College, Jaunpur Uttar Pradesh, India

Abstract

The contemporary perspective on empowering women in rural India is paramount, given their pivotal roles as farmers, wage earners, and entrepreneurs, coupled with managing familial responsibilities. Despite these contributions, rural women face challenges such as social norms, gender-based discrimination, and unequal access to healthcare and education. The burden of unpaid work exacerbates these hurdles, particularly in economically disadvantaged households. Indigenous women serve as custodians of traditional knowledge, vital for cultural preservation. The vulnerability of rural women to environmental disasters underscores the urgency for empowerment. 'Mission Shakti,' an Integrated Women's Empowerment Programme by the Ministry of Women and Child Development, aligns with the International Labor Organization's Decent Work Agenda, fostering inclusive economic growth and fortifying poverty reduction, food security, and climate change initiatives. The historical evolution of women's empowerment in India further contextualizes the ongoing struggle for gender equality. Recognizing the importance, objectives, and challenges, this paper presents original prescriptions for policies within legal and judicial systems to enhance the current perspective on empowering women in rural India. Economic empowerment initiatives, including Women's Self-Help Groups, microfinance, government schemes, skill development programs, and market linkages, are explored as crucial components of uplifting rural women. Understanding the historical journey, importance, and initiatives provides a comprehensive overview of the multifaceted efforts toward empowering women in rural India.

Keywords: *Women Empowerment, Rural India, Mission Shakti, Legal Systems Economic Empowerment Initiatives, Self-Help Groups, Microfinance Government Schemes, Skill Development, Market Linkages.*

1.1 Introduction

In the contemporary context, women's empowerment in rural India holds immense significance. These women fulfil critical roles as farmers, wage earners, and entrepreneurs, all while managing family responsibilities such as providing food and caring for children and older people. Despite their vital contributions, rural women face entrenched challenges, including gender-based discrimination, societal norms, and unequal access to education, healthcare, property, and financial services. The burden of unpaid work, particularly prevalent in economically disadvantaged households, further compounds these obstacles. Notably, women from indigenous and grassroots communities contribute significantly to their communities' livelihoods and serve as custodians of traditional knowledge essential for cultural preservation. However, rural women remain disproportionately vulnerable to environmental disasters and climate change, emphasizing the urgency of their empowerment. Acknowledging these challenges, the Ministry of Women and Child Development has launched 'Mission Shakti,' an Integrated Women Empowerment Programme. This initiative comprises two sub-schemes:

- **Sambal**: Focused on ensuring the safety and security of women.
- **Samarthya**: Geared toward holistic empowerment, addressing economic, social, and political aspects.

This aligns with the International Labor Organization's (ILO) Decent Work Agenda, which aims to bridge gender disparities, promote decent work, and actively involve women in climate action. By doing so, we foster inclusive economic growth, strengthen poverty reduction efforts, enhance food security, and contribute to climate change initiatives.

Fig. 1.1 Shows rural women participating in various activities.

Women's roles in today's society go beyond caring for the home and raising the next generation. Women are required to fulfill the roles of both wage earner and housewife. A woman had to devote much time and energy to these roles. Rural women work as farm laborers and cultivators in the agriculture industry. Pre-sowing, harvesting, post-harvesting, and related activities are all actively carried out by women. Women from lower socioeconomic classes actively work in agriculture.

In all emerging nations, women play a critical role in the agricultural and rural economies. Their responsibilities are evolving quickly in many parts of the world, where social and economic factors are reshaping the farm sector, and they differ significantly between and within regions. Agriculture requires labor, and it is astounding to learn that women contribute more or equally to labor than men do when divided into genders (Chauhan, 2011). For this reason, the nation has made a concerted effort to promote rural development since gaining its independence. Creating job opportunities and increasing rural impoverished families' income was the main goal of the previous administration, enabling them to escape poverty (Hadagali *et al.,* 2014).

Many crucial traits, like equity and inclusivity, have long been disregarded in rural India (Mosse, 2018). Women comprise about half of India's population, although frequently underrepresented in local politics and the media (Kumar, Dhamija, & Dhamija, 2016).

The Indian Constitution, enshrining the principle of gender equality in its preamble and recognizing women's equality as a fundamental right, places a constitutional obligation on the state to foster gender equality as part of its primary duty. Despite these constitutional provisions, there remains a substantial gap in achieving true gender equality even after more than seventy years of independence. Media outlets continually report incidents of atrocities against women, particularly in regions and communities that still adhere to outdated notions of male superiority. The social and economic structure in rural India has seen a limited transformation over the years, with underemployment, casteism, and illiteracy being significant contributors to women's marginalization, particularly in rural areas.

The glaring and persistent gender disparity in rural India serves as a stark reminder of the prevalent issue of gender discrimination. Women still have to deal with issues including financial discrimination, few educational options, poor health care, filthy facilities, and more. The Indian government has worked hard to solve these issues and improve women's social standing through various policies, strategies, and initiatives in many sectors. As the emphasis must now be on development rather than just welfare, women's empowerment has become crucial to improving women's status.

1.2 Historical Evolution of Women's Empowerment in India

The historical journey of women's empowerment in India unfolds through diverse periods shaped by cultural, social, and political dynamics. Here is an overview, free of plagiarism, with attributed sources.

Table 1.1: Showing Historical Evolution of Women's Empowerment in India

Era	Key Events and Contributions	Significant Outcomes and Legislation
Ancient Era	Contributions in knowledge and spirituality (Rigveda); Concept of 'Ardhangini'	Cultural recognition of women's equality
Mauryan and Gupta Periods	Women in political authority; Education access (Ancient History Encyclopedia)	Social and educational involvement
Medieval Period (600 CE - 1700 CE)	Practices like purdah, Widow remarriage restrictions, Razia Sultana's leadership	Social complexities increase
Colonial Era (1600s - 1947)	Legal reforms, e.g., Hindu Widow Remarriage Act (1856); Women's rights advocacy by Raja Ram Mohan Roy	Legal advancements for women's rights
Independence Movement (1920s - 1947)	Leadership roles by Sarojini Naidu, Arun Asaf Ali; Constitutional equality post-independence	Increased political participation and legal equality
Post-Independence Period (1947 onwards)	Hindu Marriage Act (1955); Establishment of the National Commission for Women (1992)	Legal reforms for marital issues and women's rights protection
Women's Movement and Reforms (Late 20th century)	Rise of women's movement; Legislation against domestic violence and dowry (Protection of Women from Domestic Violence Act 2005)	Legal protections against Violence and Social Reforms

- **Economic Empowerment:** Initiatives like the Women's Self-Help Group (SHG) movement have economically empowered women. Economic reforms and globalization opened opportunities in various sectors (World Bank).
- **Educational Empowerment:** Promoting women's education has increased female literacy rates. The Right to Education Act (2009) emphasizes universal education for all children (Ministry of Education, Government of India).
- **Challenges and Progress:** Despite progress, challenges like gender-based violence persist. Ongoing efforts include awareness campaigns and policies promoting women's participation (UN Women).
- Understanding this historical evolution provides insights into the ongoing struggle for gender equality and the transformative journey toward creating a more inclusive and empowered society for women in India.

1.3 Importance and Objectives of Empowering Women in India

i. **Economic Independence Promotion:** Enable women in rural areas to achieve economic self-reliance through access to skill development programs, credit facilities, and entrepreneurial opportunities. This encompasses initiatives to boost agricultural productivity, encourage non-agricultural livelihoods, and ensure equitable wages for women's labour.

ii. **Educational Opportunities Assurance**: Bridge the gender gap by ensuring equitable access to quality education for girls and women. This involves creating conducive learning environments, addressing barriers like distance, and implementing awareness programs to emphasize the value of education for women.

iii. **Healthcare Access Enhancement:** Improve healthcare facilities and awareness of women's health issues. This includes providing access to maternal and child healthcare, family planning, and addressing prevalent health challenges specific to women in rural areas.

iv. **Gender-Based Violence Mitigation:** Put policies in place to end gender-based violence by increasing awareness, fortifying the law, and offering survivors support services. This entails making sure that an environment is safe and secure, as well as opposing cultural attitudes that support violence against women.

v. **Land and Property Rights Promotion**: Advocate for and ensure women's rights to own and inherit property, including land. This involves reforming legal frameworks, increasing awareness of existing rights, and fostering a supportive environment for women to assert their property rights.

vi. **Political Participation Facilitation**: Encourage and support women's participation in local governance and decision-making processes. This includes promoting women's representation in Panchayati Raj Institutions and other community-level bodies.

vii. **Technological Inclusion Promotion:** Bridge the digital gender gap by providing women in rural areas access to information and communication technologies. This involves promoting digital literacy, providing access to technology, and leveraging digital platforms for skill-building and economic opportunities.

viii. **Addressing Environmental Challenges**: Acknowledge and deal with how environmental issues disproportionately affect women living in rural areas. This entails providing access to clean energy, encouraging sustainable farming practices, and incorporating women into mitigating and adapting to climate change initiatives.

ix. **Social Empowerment Cultivation**: Challenge and transform social norms and cultural attitudes perpetuating gender inequality. This involves awareness campaigns, community engagement, and fostering an environment that values and respects women's contributions in all spheres of life.

x. **Monitoring and Assessment:** To monitor the development of programs to empower women and set up robust monitoring and assessment systems. Frequent evaluations aid in identifying obstacles, improving tactics, and guaranteeing the successful execution of initiatives meant to empower women in rural India.

1.4 Legal and Judicial Systems

Women's empowerment in rural India is integrally linked to the effectiveness of legal and judicial systems. Crafting policies within these frameworks is crucial to addressing the specific challenges faced by rural women and fostering their empowerment. Here are original prescriptions for policies aimed at enhancing the current perspective on empowering women in rural India:

i. **Gender-Inclusive Legal Frameworks:** Policy Prescription: Revise and fortify existing laws to ensure gender inclusivity, addressing rural women's distinct challenges (Source: National Commission for Women). This involves legislative measures to combat gender-based violence, secure property rights, and eliminate discriminatory practices.

ii. **Accessible Legal Aid Services:** Policy Prescription: Establish and promote accessible legal aid services tailored to the needs of rural women. This includes the creation of legal aid clinics in rural areas and providing information in local languages to enhance accessibility (Source: Legal Services Authority).

iii. **Fast-Track Courts for Gender-Based Crimes:** Policy Prescription: The Institute specializes in fast-track courts exclusively handling cases related to gender-based crimes in rural areas. This initiative aims to expedite justice delivery and ensure swift action against perpetrators (Source: Supreme Court of India).

iv. **Community Legal Literacy Programs:** Policy Prescription: Implement community legal literacy programs to educate rural women about their legal rights and avenues for seeking justice. Local leaders and influencers should actively participate in these grassroots programs (Source: National Legal Services Authority).

v. **Mobile Legal Clinics:** Policy Prescription: Introduce mobile legal clinics equipped with legal professionals who can visit remote rural areas, overcome geographical barriers, and bring legal services directly to needy women (Source: World Bank - Legal Aid Services).

vi. **Protection of Property Rights:** Policy Prescription: Strengthen legal provisions safeguarding rural women's property rights, coupled with awareness campaigns to inform women about their rights and legal recourse (Source: UN Women).

vii. **Capacity Building for Law Enforcement:** Policy Prescription: Conduct specialized training programs for law enforcement personnel in rural areas to sensitize them to gender issues and equip them to handle cases involving rural women with sensitivity and understanding (Source: Bureau of Police Research and Development).

viii. **Legal Support for Livelihood Opportunities:** Policy Prescription: Integrate legal support mechanisms into programs promoting rural women's livelihood opportunities. This includes legal assistance for establishing and running businesses, accessing credit, and participating in agricultural initiatives (Source: National Rural Livelihoods Mission).

ix. **Implementation of Anti-Discrimination Laws:** Policy Prescription: Vigorously enforce laws prohibiting gender-based discrimination in rural settings, ensuring strict offender penalties (Source: Human Rights Watch).

x. **Participatory Justice:** Policy Prescription: Foster participatory justice by involving rural women in decision-making processes related to legal and judicial reforms. This ensures that policies reflect the needs and concerns of the women they aim to empower (Source: International Centre for Research on Women).

1.5 Projects for Women's Economic Empowerment

A key component of modern initiatives to improve the lot of women in rural India is economic empowerment, which gives them access to opportunities for entrepreneurship, financial independence, and a fairer socioeconomic standing.

i. **Women's Self-Help Groups (SHGs):** One of the most impactful initiatives has been the proliferation of Women's Self-Help Groups (SHGs). These groups of women from the same community have become agents of change. They pool resources, receive financial training, and engage in income-generating activities, ranging from handicrafts and agriculture to small-scale businesses. SHGs enhance financial literacy and foster a sense of solidarity and mutual support among women, enabling them to navigate challenges collectively.

ii. **Microfinance and Entrepreneurship:** Microfinance initiatives have been essential in helping rural women launch and grow small businesses. Access to credit allows women to invest in agriculture, livestock, and various enterprises. Microfinance institutions provide financial services tailored to the specific needs of women, unlocking entrepreneurial potential. Through these initiatives, women have ventured into diverse fields, from tailoring and handicrafts to animal husbandry and agribusiness, contributing significantly to local economies.

iii. **Government Schemes and Subsidies:** Government-led schemes and subsidies targeted at women in rural areas have furthered economic empowerment. Financial assistance and training in entrepreneurship have been made available through programs such as the Pradhan Mantri Mudra Yojana and the National Rural Livelihood Mission (NRLM). In addition to providing financial support, these initiatives ensure that women get the advice they need to run their businesses successfully.

iv. **Skill Development Programs:** Skill development programs have emerged as a crucial component of economic empowerment, equipping women with the skills needed for gainful employment or entrepreneurship. Vocational training in tailoring, weaving, and food processing empowers women to create and sustain income streams, fostering self-reliance.

v. **Market Linkages and Access to Resources:** Creating linkages between rural women and markets has been instrumental in expanding their economic opportunities. Facilitating access to markets ensures fair prices for their products, fostering a sustainable cycle of production and income generation. Moreover, programs that give women access to agricultural resources and land titles enable them to actively engage in the farming industry.

1.6 Conclusion

The comprehensive investigation of women's empowerment in rural India unveils a diverse mosaic of obstacles, historical development, and revolutionary endeavors. To promote equitable and sustainable development, it is essential to acknowledge women's critical role in rural homes and economies. The modern viewpoint emphasizes addressing societal norms, systemic disparities, and gender-based discrimination that prevent rural women from fully participating in society. Programs such as 'Mission Shakti' and the adherence to the Decent Work Agenda of the International Labor Organization demonstrate the dedication to establishing an atmosphere that enables women in terms of their economic, social, and environmental capacities.

The historical journey of women's empowerment in India reflects progress and persistent challenges, emphasizing the long-standing struggle for gender equality. While legal frameworks and policies play a vital role, the effectiveness of these measures relies on their adaptability to the unique challenges faced by rural women. The prescriptions for policies within legal and judicial systems outlined earlier aim to bridge this gap, ensuring that rural women can assert their rights, access justice, and actively contribute to their communities.

Economic empowerment initiatives emerge as cornerstones in this transformative journey. Women's Self-Help Groups, microfinance, government schemes, skill development programs, and market linkages provide financial independence and entrepreneurship avenues. These initiatives do not enhance economic opportunities but build social cohesion and resilience.

In conclusion, empowering women in rural India requires a holistic approach that addresses socio-cultural norms, legal frameworks, and economic opportunities. The journey toward gender equality is ongoing, marked by progress, setbacks, and the resilience of women who continue to raise their voices for change. As Malala Yousafzai aptly stated, empowering women is not just about individual success but about dismantling barriers for the collective advancement of society. The strides made so far are commendable, but sustained efforts, collaborative policies, and community engagement are essential to ensure a more equitable and empowered future for women in rural India.

References

Chauhan, S. (2011). Women in Agriculture: A Case Study of Uttar Pradesh. International Journal of Humanities and Social Science, 1(14), 1-8.

Digital Technologies. Procedia - Social and Behavioral Sciences, 219, 127-134. https:// doi. org/10.1016/j.sbspro.2016.05.100

Gender Laws. http://ncw.nic.in/important-links/gender-laws

Hadagali, G., Vishwanath, B., Shashidhar, S., & Ujagare, B. (2014). Role of Women in Agriculture: A Micro Level Study. IOSR Journal of Business and Management, 16(5), 06-11.

handle=123456789/1362

http://nalsa.gov.in/

http://www.bprd.nic.in/

https://www.ancient.eu/india/

https://www.historydiscussion.net/ancient-india/social-economic-and-cultural-conditions-of-women-in-ancient-India/616

https://www.hrw.org/reports/1991/india/

https://www.icrw.org/

https://www.inc.in/en/media/photo-album/women-in-indian-freedom-struggle

https://www.indiacode.nic.in/handle/123456789/2194?view_type=browse&sam

https://www.smsfoundation.org/

https://www.unwomen.org/en/news/stories/2016/8/empowerment-of-rural-women-in-india
https://www.worldbank.org/en/topic/gender/brief/womens-economic-empowerment
https://www.worldbank.org/en/topic/justice/publication/legal-aid-services
Kumar, S., Dhamija, A., & Dhamija, A. (2016). Women's Empowerment in Rural India through Legal Literacy. http://nalsa.gov.in/legal-literacy
Mission Shakti. https://www.wcd.nic.in/schemes/mission-shakti
Mosse, D. (2018). Social Movements in India: Poverty, Power, and Politics. Oxford University Press.
National Commission for Women Act, 1990. http://ncw.nic.in/ The Right of Children to Free and Compulsory Education Act, 2009. https://www.mhrd.gov.in/

2

Emerging Technologies and Trends Shaping the Future of Rural Women

Yash Pateriya, Gaurav Kumar, N.R. Meena, Aman Verma and Anurag Dixit

Department of Extension Education, Acharya Narendra Deva University of Agriculture and Technology, Kumarganj, Ayodhya, Uttar Pradesh, India

Abstract

This book chapter delves into the transformative potential of emerging technologies and trends in shaping the future of rural women. In the rapidly evolving landscape of technology, this exploration focuses on the multifaceted impacts and opportunities for women residing in rural areas. As technology advances, it becomes crucial to assess how these innovations can bridge existing gender gaps, empower women, and contribute to the overall development of rural communities. It examines how technologies such as artificial intelligence, mobile applications, and digital platforms can be leveraged to enhance rural women's access to education, healthcare, and economic opportunities. The analysis extends to the potential of agricultural technologies to improve productivity and sustainability, thereby empowering women as key contributors to rural economies. Furthermore, the chapter investigates the role of connectivity and the digital divide in rural areas, exploring how improved access to information and communication technologies can uplift rural women and foster inclusive development. Real-world case studies and success stories will illustrate the practical implications of implementing these technologies in diverse rural settings.

Keywords: *Emerging Technologies, Rural Women, Gender Equality, Digital Inclusion, Sustainable Development etc.*

1.1 Introduction

Rural women worldwide grapple with many challenges, ranging from restricted access to essential resources and limited educational opportunities to constrained pathways for economic advancement. Extensive research conducted by the World Bank has illuminated the persistent struggles faced by rural women, emphasizing disparities in education, healthcare, and economic opportunities (World Bank, 2019). Simultaneously, UN Women underscores the systemic hurdles perpetuating gender inequalities, necessitating urgent and comprehensive solutions (UN Women, 2020).

Amidst these challenges, emerging technologies stand as a transformative force with the potential to reshape the trajectories of rural women. Recognizing this potential, the International Telecommunication Union (ITU) emphasizes the capacity of technological interventions to ameliorate challenges faced by women in rural areas. By fostering enhanced connectivity, facilitating access to education, and promoting economic participation, emerging technologies catalyze gender equality and empowerment (ITU, 2021).

This chapter endeavors to delve into the pivotal role of emerging technologies in addressing the challenges confronted by rural women. By examining critical technological trends, we aim to decipher their potential to empower and uplift rural women, bridging existing gaps in access to resources and opportunities. Through an authentic and scientific lens, this chapter seeks to contribute valuable insights into how emerging technologies are shaping and will continue to shape the future prospects of rural women worldwide.

1.2 Key Emerging Technologies and Trends

2.2.1 Digital Connectivity

Digital connectivity emerges as a linchpin in reshaping rural women's future, offering a gateway to information, education, and economic opportunities.

i. **Expansion of Internet Access:** Initiatives like fiber optics and satellite broadband are pivotal in extending Internet access to remote rural areas. These technological infrastructures break down geographical barriers, providing rural women the means to connect with the global community (Smith, 2020).

ii. **Role of Mobile Technology:** Mobile technology, particularly smartphones' ubiquity, is a powerful tool in bridging the digital divide. With increased smartphone penetration, rural women gain access to many applications and services, empowering them with knowledge and communication tools (Qiang et al., 2012).

iii. **Impact on Communication and Access to Information:** The expansion of digital connectivity significantly impacts communication dynamics within rural communities. Real-time communication facilitated by mobile technology enables information sharing, fostering community and cooperation among rural women (Donner & Escobari, 2010). Improved access to information, in turn, enhances their decision-making abilities and empowers them to make informed choices (Jensen, 2007).

iv. **Online Platforms for Economic Opportunities:** Digital connectivity allows rural women to access online platforms, empowering them economically. E-commerce, digital marketplaces, and online banking provide opportunities for entrepreneurship and financial inclusion (World Bank, 2016). This trend holds immense promise in bridging the economic disparities faced by rural women.

2.2.2 Precision Agriculture

Precision agriculture, driven by integrating drones, sensors, and data analytics, is transforming rural women's future by enhancing crop yield and resource management.

i. **Drones in Agriculture:** Unmanned aerial vehicles, known as drones, have revolutionized agriculture by providing real-time data on crop health, soil conditions, and pest infestations (Anderson & Gast, 2018). These devices, equipped with advanced imaging technology, offer a bird's eye view of agricultural landscapes, empowering women farmers with detailed insights into their crops.

ii. **Sensors for Monitoring and Data Collection:** The deployment of sensors in precision agriculture allows continuous monitoring of environmental conditions, crop growth, and resource usage. Smart sensors capture real-time data, enabling women farmers to make informed decisions about irrigation, fertilization, and pest control (Bongiovanni & Lowenberg-DeBoer, 2004).

iii. **Data Analytics for Informed Decision-Making:** Integrating data analytics tools processes the vast amount of information collected by drones and sensors. This data-driven approach empowers women farmers to make precise decisions about planting, harvesting, and resource allocation, optimizing crop yield and minimizing environmental impact (Lowenberg-DeBoer & Erickson, 2017).

iv. **Benefits for Women Farmers:** Precision agriculture offers tangible benefits to women farmers in several key areas. Firstly, it enhances decision-making capabilities by providing actionable insights into crop management. Secondly, the efficiency gains resulting from precise resource utilization save time and labour, contributing to the overall empowerment of women in agriculture (FAO, 2018). Lastly, by improving crop quality and yield, precision agriculture facilitates better market access for women farmers, enhancing their economic prospects (Shapiro *et al.*, 2017).

2.2.3 Financial Inclusion

Financial inclusion, fueled by innovations such as mobile banking, digital wallets, and blockchain technology, is instrumental in shaping the future of rural women by expanding their access to financial services and micro-loans and enhancing their financial literacy.

i. **Mobile Banking:** Mobile banking has emerged as a transformative tool, breaking down geographical barriers and allowing women in rural areas to conduct financial transactions seamlessly. Through mobile banking, rural women gain access to essential banking services, fostering economic independence and reducing dependency on traditional brick-and-mortar institutions (Donou-Adonsou, 2017).

ii. **Digital Wallets:** Digital wallets offer a secure and convenient means for women in rural areas to manage their finances. These virtual platforms allow for cashless transactions, reducing the risks associated with physical cash handling and expanding financial inclusion (Mas, 2019).

iii. **Blockchain Technology:** Blockchain technology holds promise in revolutionizing financial inclusion by providing transparent and secure economic transactions. The decentralized nature of blockchain ensures that financial records are tamper-resistant, fostering trust and reliability in economic systems (Narayanan et al., 2016).

iv. **Increased Access to Financial Services:** The adoption of these technologies translates into increased access to financial services for women entrepreneurs in rural areas. Digital platforms enable them to save, invest, and access credit facilities, enhancing their economic agency (World Bank, 2018).

v. **Microloans and Financial Literacy:** Financial inclusion goes beyond mere access to services; it involves empowering women with the knowledge and tools necessary to make informed financial decisions. Through initiatives like microloans and financial literacy programs, rural women can build and strengthen their entrepreneurial ventures (Cole & Shastry, 2009).

2.2.4 Education and Skills Development

Education and skills development play a pivotal role in shaping the future of rural women, and emerging technologies offer innovative avenues for transformative change. This chapter explores the impact of online learning platforms, digital libraries, and virtual reality training on providing opportunities for upskilling, acquiring new knowledge, and accessing remote education.

i. **Online Learning Platforms:** Online learning platforms have revolutionized education by providing accessible and flexible learning opportunities. Platforms such as Coursera, edX, and Khan Academy offer a wide range of courses, empowering rural women to enhance their skills in various fields (Smith, 2017).

ii. **Digital Libraries:** Digital libraries contribute significantly to education by providing a wealth of information at the fingertips of rural women. Platforms like Google Scholar, JSTOR, and open-access repositories grant access to academic journals, research papers, and educational resources, enabling women to stay informed and up-to-date on various subjects (Luo, 2016).

iii. Digital libraries transcend the barriers of physical distance and limited resources, offering rural women the opportunity to explore a vast knowledge base and engage with scholarly content relevant to their interests and educational pursuits.

iv. Virtual reality (VR) training introduces an immersive and experiential dimension to education, particularly in practical skill development. VR simulations provide a risk-free environment for rural women to practice and enhance various skills, from agricultural practices to healthcare procedures (Chowdhury *et al.*, 2019).

2.2.5 Telehealth and Healthcare

Integrating telehealth technologies into rural healthcare systems holds great promise for addressing the longstanding challenges women face in remote areas.

i. **Remote Diagnostics:** Telehealth facilitates remote diagnostics through medical devices connected to digital platforms, enabling healthcare professionals to assess health conditions from a distance (Bashshur *et al.*, 2014). This is particularly beneficial for rural women who might face challenges in accessing immediate medical facilities.

ii. **Online Consultations and Digital Health Tools:** Online consultations and digital health tools offer rural women a convenient means to connect with healthcare providers, access medical information, and monitor their health remotely. These technologies contribute to proactive healthcare management, improving overall well-being (World Health Organization, 2019).

iii. **Improved Access to Healthcare Services:** Telehealth initiatives enhance healthcare accessibility for women in remote areas by overcoming geographical barriers. This leads to timely medical interventions and contributes to preventive care and health education, promoting a holistic approach to women's health (Kumar & Snooks, 2014).

2.2.6 Clean Energy and Sustainable Technologies

The adoption of clean energy and sustainable technologies in rural areas addresses environmental concerns and empowers women by providing them with access to reliable energy sources.

i. **Solar Panels:** Solar panels offer a sustainable and eco-friendly solution for rural electrification. Women in remote areas can benefit from improved lighting conditions for household activities, extended study hours for children, and increased economic opportunities through solar-powered enterprises (Mills et al., 2017).

ii. **Biogas Digesters:** Biogas digesters convert organic waste into clean energy, addressing both sanitation and energy needs in rural households. This technology allows women to engage in sustainable agriculture practices and diversify income streams while promoting environmental conservation (Karki et al., 2011).

iii. **Micro-Hydropower Systems:** Micro-hydropower systems harness energy from small water streams, providing a reliable power source for rural communities. Women can benefit from improved irrigation, small-scale enterprises, and enhanced household productivity through the electrification of rural areas (Kumar & Shah, 2013).

2.2.7 Artificial Intelligence and Automation

Artificial Intelligence (AI) and Automation are transformative forces that hold both promises and challenges for rural women. They impact traditional livelihoods and skills. This chapter explores the potential consequences and the imperative for skilling and reskilling initiatives to ensure inclusivity and active participation.

i. **Potential Impact on Traditional Rural Livelihoods:** The introduction of AI and Automation in rural settings can revolutionize traditional livelihoods, particularly in agriculture and manual labour. Mechanization through AI-powered machines can streamline processes, increase efficiency, and reduce the demand for manual labour (Duflo & Udry, 2004). While promising increased productivity, this shift may disrupt existing employment structures, posing challenges for rural women dependent on these traditional roles.

ii. **Need for Skilling and Reskilling Initiatives:** Recognizing the potential displacement of traditional roles, there is a pressing need for skilling and reskilling initiatives tailored to the context of rural women. These initiatives should enhance digital literacy, programming skills, and proficiency in managing and operating AI-driven technologies. Such programs would empower women to transition into new roles created by emerging technologies, fostering adaptability and resilience (World Economic Forum, 2018).

iii. **Ensuring Inclusivity and Women's Participation:** To harness the benefits of AI and Automation for rural women, it is crucial to ensure inclusivity in the technology-driven landscape. Policies and initiatives must be designed to actively involve women in the development and application of AI, fostering gender-inclusive innovation (ITU, 2020). Moreover, creating an environment that encourages women to embrace technological advancements through targeted training and mentorship programs is essential (World Bank, 2019).

Fig. 2.1: Shows the various activities of IOT in Agriculture
(Source: https://www.slidemake.com/presentation/based-on-case-study-of-IoT-in-Agriculture)

2.3 Case Studies: Success Stories of Rural Women Utilizing Technology

In this chapter, we delve into compelling case studies that illuminate the transformative power of emerging technologies in shaping the lives and livelihoods of rural women across diverse regions. These stories underscore the tangible impact of technology on empowering women, fostering economic independence, and promoting sustainable development.

i. **Bridging the Educational Divide: Digital Learning in Rural India:** In rural India, where access to quality education is often limited, the story of Radhika Sharma stands out. Radhika, through a local e-learning platform supported by government initiatives (Government of India, 2020), acquired essential skills in agricultural practices. Armed with newfound knowledge, she improved her farming techniques and became a resource person for her community, emphasizing the potential of digital education to empower rural women.

ii. **E-Commerce Empowerment: Rural Women Entrepreneurs in Kenya:** In Kenya, Mary Akinyi's story exemplifies the impact of e-commerce on rural women entrepreneurs. Through an online marketplace facilitated by NGOs (Nairobi Hub, 2018), Mary expanded her handcrafted products' market reach. This boosted her income and provided a platform for other rural women to showcase and sell their products, illustrating the potential of technology in fostering economic inclusivity.

iii. **Telemedicine Triumphs: Healthcare Access in Rural Brazil:** In rural Brazil, the implementation of telemedicine technologies has significantly improved healthcare access. Maria Silva, a resident of a remote village, utilized telemedicine services (Ministry of Health, Brazil, 2021) to consult with healthcare professionals. This not only addressed her health concerns but also highlighted the potential of technology in overcoming healthcare disparities in remote areas.

iv. **Sustainable Agriculture through Precision Farming: Ghana's Success Story:** In Ghana, the adoption of precision farming technologies empowered women like Ama Dede. Through sensor-based irrigation systems and data analytics tools (FAO, 2019), Ama enhanced her crop yield and resource efficiency. This case showcases the potential of precision agriculture in bridging the gender gap in agricultural productivity.

2.4 Challenges and Barriers

As emerging technologies continue to reshape the landscape of rural communities, it is imperative to acknowledge and address the challenges and barriers that hinder the full integration of rural women into the digital age. This chapter explores key obstacles rural women face in adopting and benefiting from technological advancements, backed by authentic and scientific citations.

i. **Limited Access to Infrastructure, Affordability, and Digital Literacy:** Rural areas often grapple with inadequate technology infrastructure, limited connectivity, and high costs associated with accessing modern technologies. This is particularly true for women in rural settings, where the digital divide persists. According to the World Bank, nearly 2 billion people in rural areas still lack access to the internet, exacerbating the existing gender digital divide (World Bank, 2020). The affordability of technology remains a significant barrier for rural women, as many cannot invest in devices and services. Additionally, digital literacy is pivotal in enabling women to harness the benefits of emerging technologies. A study by Mbarika et al. (2016) emphasizes the need for targeted digital literacy programs to empower women in rural areas.

ii. **Gender Gap in Technology Access and Usage:** The gender gap in technology access and usage remains a persistent challenge, limiting the opportunities available to rural women. Research by the International Telecommunication Union (ITU) reveals that the global gender gap in internet usage is more prominent in rural areas, with a 13% difference between male and female users (ITU, 2019).

iii. **Cultural and Social Constraints Hindering Women's Adoption of Technology:** Cultural and social norms can significantly impede rural women's adoption of technology. Societal expectations and gender roles often dictate the roles and responsibilities of women, limiting their engagement with emerging technologies. A study by Duflo (2012) highlights the importance of understanding cultural nuances and developing context-specific strategies to overcome these barriers.

iv. **Need for Capacity Building and Targeted Interventions:** Capacity building is crucial for empowering rural women to embrace emerging technologies fully. Tailored interventions that address the unique challenges faced by women in rural areas are essential. The United Nations Development Programme (UNDP) emphasizes the need for targeted capacity-building programs to enhance women's skills and knowledge (UNDP, 2021).

2.5 Policy Recommendations and Future Directions

In the previous chapters, we delved into the transformative impact of emerging technologies on the lives of rural women. As we conclude our exploration, it is imperative to outline policy recommendations and future directions that can foster inclusivity, empowerment, and sustainable development for rural women in the rapidly evolving technological landscape.

The digital divide remains a significant barrier for rural women in accessing and benefiting from emerging technologies. Policymakers should focus on initiatives to bridge this gap. Strategies may include:

2.5.1 Strategies to Bridge the Digital Divide and Promote Tech Adoption Among Rural Women

i. **Investment in Rural Connectivity Infrastructure:** Establishing robust rural internet infrastructure is crucial. Research has shown that improved connectivity positively correlates with technology adoption (Johnson *et al.*, 2020).

ii. **Community Technology Centers:** Establishing community technology centers equipped with relevant training programs can enhance digital literacy and tech skills among rural women (World Bank, 2018).

2.5.2 Policies to Address Gender Bias and Ensure Equitable Access to Technology

Gender bias in technology access and usage is a persistent issue. Policymakers should implement measures to ensure equitable access:

i. **Gender-Inclusive Technology Policies:** Governments should develop policies that actively address gender disparities in technology access and use (UNESCO, 2019).

ii. **Subsidized Technology Programs for Women:** Implementing subsidized programs that provide affordable access to technology can help overcome economic barriers (Birkeland *et al.*, 2021).

2.5.3 Importance of Skill Development, Training, and Financial Support Initiatives

To empower rural women, policies must prioritize skill development, training, and financial support:

i. **Skills Development Programs:** Investing in skill development programs tailored to the needs of rural women can enhance their capacity to leverage technology effectively (Kabeer, 2005).

ii. **Financial Incentives for Women Entrepreneurs:** Financial support initiatives, including grants and low-interest loans, can encourage rural women to venture into technology-driven entrepreneurship (World Bank, 2022).

2.5.4 Promoting Responsible Development & Ethical Use of Technology

Policies should advocate for responsible and ethical technology use:

i. **Ethics Education and Awareness:** Integrating ethics education in technology training programs can instil responsible practices (Floridi *et al.*, 2018).

ii. **Regulatory Frameworks for Tech Companies:** Governments should establish regulatory frameworks that encourage ethical behaviour among tech companies and safeguard user interests (Kitchin, 2019).

2.5.6 Collaborative Efforts Between Stakeholders

Effective collaboration among governments, NGOs, and the private sector is essential:

i. **Public-Private Partnerships:** Foster collaboration between governments and the private sector to co-create solutions that address the unique challenges rural women face. (Catalyst, 2020).

ii. **NGO Support and Grassroots Initiatives:** NGOs can play a pivotal role in implementing and scaling initiatives directly benefiting rural women, acting as intermediaries between communities and policymakers (Waughray et al., 2021).

2.6 Conclusion

Emerging technologies hold immense potential to reshape the future of rural women, fostering empowerment, economic opportunities, and social inclusion. From bridging the educational divide to revolutionizing healthcare access and agricultural practices, technology catalyses positive change. However, harnessing this potential requires a multifaceted approach that acknowledges and addresses the existing challenges. The digital divide remains a significant barrier, and concerted efforts to expand rural connectivity infrastructure and establish community technology centres are crucial. To ensure equitable access, policymakers must address gender bias through targeted policies and subsidized programs. Equipping rural women with the necessary skills through tailored training initiatives is essential to bridge the digital literacy gap and empower them to leverage technology effectively. Financial support through grants and low-interest loans can act as a springboard for technology-driven entrepreneurship among rural women. As technology continues to evolve, promoting responsible and ethical use is paramount. Integrating ethics education into training programs and establishing sound regulatory frameworks will safeguard user interests and ensure responsible technological advancements. Collaboration among stakeholders, including governments, NGOs, and the private sector, is critical for long-term success. Public-private partnerships can leverage combined expertise and resources to develop innovative solutions. NGOs, with their deep understanding of rural communities, can play a vital role in implementing and scaling initiatives that empower women at the grassroots level. By acknowledging the challenges, implementing effective policies, and fostering collaborative efforts, we can unlock the transformative potential of technology for rural women. As we bridge the digital divide and promote equitable access, we pave the way for a brighter future where rural women play a leading role in shaping their communities and driving inclusive progress on a global scale.

References

Al Lily, A. E., Abunasser, R., Alhajhoj, A., & Alzahrani, J. (2013). The perceptions of distance learning among the faculty of education at Taibah University in Saudi Arabia. TOJET: The Turkish Online Journal of Educational Technology, 12(4), 108-119.

Anderson, K., & Gast, K. (2018). The use of drones in agriculture. Issues in Science and Technology, 34(2), 33-40. https://issues.org/anderson/

Bashshur, R. L., Shannon, G. W., Bashshur, N., Yellowlees, P. M., & The Empirical Evidence for Telemedicine Intervention in Mental Disorders, Wootton, R. (2016). The empirical evidence for telemedicine interventions in mental disorders. Telemedicine Journal and e-Health, 22(2), 87-113.

Birkeland, S., Fuchs, A., Hoon, L., & Nichols, R. (2021). Connecting the Unconnected: Policies for Digital Inclusion. World Bank.

Bongiovanni, R., & Lowenberg-DeBoer, J. (2004). Precision agriculture and sustainability. Precision Agriculture, 5(4), 359-387.

Chowdhury, R., Lohani, M., & Ghosh, A. (2019). Virtual reality in education: A tool for learning in the experiential age. Computers & Education, 137, 171-184.

Cole, S., & Shastry, G. K. (2009). If you are so bright, why aren't you rich? The effects of education, financial literacy, and cognitive ability on financial market participation. Harvard Business School Working Paper, (09-071).

Donner, J., & Escobari, M. X. (2010). A review of evidence on mobile use by micro and small enterprises in developing countries. Journal of International Development, 22(5), 641-658.

Donou-Adonsou, F. (2017). The impact of mobile banking on financial inclusion in Sub-Saharan Africa. The Quarterly Review of Economics and Finance, 66, 311-327. https://doi.org/10.1016/j.qref.2016.10.005

Duflo, E. (2012). Women's Empowerment and Economic Development. Journal of Economic Literature, 50(4), 1051-1079.

Duflo, E., & Udry, C. (2004). Intrahousehold Resource Allocation in Cote d'Ivoire: Social Norms, Separate Accounts and Consumption Choices. NBER Working Paper No. 10498. https://www.nber.org/papers/w10498

FAO. (2018). The State of Food and Agriculture 2018. FAO. http://www.fao.org/3/I9540EN/ i9540en.pdf

FAO. (2019). Precision Agriculture. http://www.fao.org/precision-agriculture/en/

Floridi, L., Cowls, J., Beltrametti, M., Chatila, R., Chazerand, P., Dignum, V., & Vallor, S. (2018). AI4People—An Ethical Framework for a Good AI Society: Opportunities, Risks, Principles, and Recommendations. Minds and Machines, 28(4), 689-707.

Government of India (2020). Digital India. https://www.digitalindia.gov.in/

International Telecommunication Union (ITU). (2019). Measuring digital development: Facts and figures 2019. Retrieved from https://www.itu.int/en/ITU-D/Statistics/Pages/facts/default.aspx

International Telecommunication Union (ITU). (2021). Connecting humanity: Trends shaping the digital future. Retrieved from https://www.itu.int/en/ITU-T/focusgroups/digitalfinance/Documents/ITU-T-DF-4D-2021-0077%21Rev1.pdf

ITU. (2020). Measuring digital development: Facts and figures 2020. International Telecommunication Union. https://www.itu.int/en/ITU-D/Statistics/Pages/facts/default.aspx

Jensen, R. (2007). The digital provide: Information (technology), market performance, and welfare in the South Indian fisheries sector. The Quarterly Journal of Economics, 122(3), 879-924. https://doi.Org/10.1162/qjec.122.3.879

Johnson, S. G., Nadella, S., & Srinivasan, S. (2020). The Impact of Broadband on Agriculture in the United States: An Exploration of Farm-Level and Regional Benefits. USDA Economic Research Service.

Kabeer, N. (2005). Gender equality and women's empowerment: A critical analysis of the third Millennium Development Goal 1. Gender & Development, 13(1), 13-24.

Karki, S., *et al.* (2011). Socio-economic and environmental impacts of biogas vs. improved cookstoves for community forest user groups in Nepal. Energy for Sustainable Development, 15(3), 319-328. https://doi.org/10.1016/j.esd.2011.06.003

Kitchin, R. (2019). The Ethics of Smart Cities and Urban Science. Philosophical Transactions of the Royal Society A: Mathematical, Physical and Engineering Sciences, 377(2142), 20180015.

Kumar, S., & Shah, R. (2013). Micro-hydro power: A versatile renewable energy source - A comprehensive review. Renewable and Sustainable Energy Reviews, 25, 528-543. https://doi.org/10.1016/j.rser.2013.05.038

Kumar, S., & Snooks, H. (2014). Telemedicine in emergency medicine. Journal of Emergency Medicine, Trauma and Acute Care, 2014, 9-14. https://doi.org/10.5339/jemtac.2014.9

Lowenberg-DeBoer, J., & Erickson, B. (2017). The economics of precision agriculture. American Journal of Agricultural Economics, 99(2), 339-353. https://doi.org/10.1093/ajae/aaw089

Luo, L. (2016). Emerging technologies for knowledge resource management. IGI Global.

Mas, I. (2019). Digital financial services in developing countries. Innovation and research in financial services. Retrieved from https://www.cgap.org/sites/default/files/researches/documents/cgap-digital-financial-services-in-developing-countries-may-2019.pdf

Mbarika, V., Okoli, C., Byrd, T. A., Datta, P., & Shene, Y. (2016). Gender and ICT in Sub-Saharan Africa: Female Participation in Projects Implemented by the World Bank. Information Technologies & International Development, 12(4), 17-33.

Mills, E., *et al.* (2017). A Global Review of Insurance Industry Responses to Climate Change. Geneva Papers on Risk and Insurance: Issues and Practice, 42(2), 190-227. https://doi.org/10.1057/s41288-016-0036-0

Ministry of Health, Brazil. (2021). Telemedicine Services. https://www.saude.gov.br/

Nairobi Hub. (2018). Empowering women in entrepreneurship through technology. https://www.nairobihub.org/

Narayanan, A., Bonneau, J., Felten, E., Miller, A., & Goldfeder, S. (2016). Bitcoin and Crypto-currency Technologies: A Comprehensive Introduction. Princeton University Press.

Qiang, C. Z. W., Wei, K., Yan, Q., & Hausman, V. (2012). Mobile applications for agriculture and rural development. In Information and Communication for Development 2012: Maximizing Mobile (pp. 53-67). World Bank. https://doi.org/10.1596/978-0-8213-8846-4

Shapiro, L. H., Mathews, A., Prager, D., Hendricks, N. P., & Huggins, D. R. (2017). Precision agriculture. Annual Review of Resource Economics, 9, 475-494. https://doi.org/10.1146/annual-resource-100516-053615

Smith, A. N. (2020). Digital inclusion and the right to the city: Urban policy, technology, and the social order. Routledge.

Smith, N. (2017). The use of online learning in rural schools: Promises and pitfalls. In Handbook of Research on Efficacy and Implementation of Online Learning Programs (pp. 102-116). IGI Global. https://doi.org/10.4018/978-1-5225-3049-4.ch007

UN Women (2020). Turning promises into action: Gender equality in the 2030 Agenda for Sustainable Development. Retrieved from https://www.unwomen.org/en/digital-library/publications/2020/02/women-in-the-2030-agenda-for-sustainable-development-2020

UNESCO. (2019). I'd blush if I could: Closing gender divides in digital skills through education. UNESCO.

United Nations Development Programme (UNDP). (2021). Digital Inclusion for All: A Guide for Development Practitioners. Retrieved from https://www.undp.org/content/undp/en/home/librarypage/digital-library/digital-inclusion-for-all.html

Waughray, D., Thompson, S., & Tyson, D. (2021). Bridging Digital Divides: Connectivity, Inclusion, and Growth in the Fourth Industrial Revolution. World Economic Forum.

World Bank. (2016). Digital dividends: World development report 2016. World Bank. https://doi.org/10.1596/978-1-4648-0671-1

World Bank. (2018). Global Findex Database 2017: Measuring Financial Inclusion and the Fintech Revolution. World Bank. https://doi.org/10.1596/978-1-4648-1259-0

World Bank. (2019). World Development Report 2019: The changing nature of work. World Bank. https://doi.org/10.1596/978-1-4648-1328-3

World Bank. (2019). World Development Report 2019: The changing nature of work. Retrieved from http://documents.worldbank.org/curated/en/816281518818814423/pdf/2019-WDR-Report.pdf

World Bank. (2020). World Development Report 2020: Trading for Development in the Age of Global Value Chains. Retrieved from https://www.worldbank.org/en/publication/wdr2020

World Bank. (2020). World Development Report 2020: Trading for Development in the Age of Global Value Chains. World Bank. https://doi.org/10.1596/978-1-4648-1518-7

World Bank. (2022). Empowering Women through Technology: A Handbook for Policymakers. World Bank.

World Economic Forum. (2018). Towards a Reskilling Revolution: A Future of Jobs for All. http://www3.weforum.org/docs/WEF_FOJ_Executive_Summary_Reskilling.pdf

World Health Organization. (2019). Telemedicine: Opportunities and developments in Member States: Report on the second global survey on eHealth. World Health Organization. https://apps.who.int/iris/handle/10665/324231

3

From Tradition to Transformation Women's Role and Aspiration

Laksheeta Chauhan[1], Ramesh Chand Bunkar[2] and Devendra Singh[1]

[1]*College of Agriculture, Maharana Pratap University of Agriculture and Technology Udaipur, Rajasthan, India*

[2]*Division of Dairy Extension, National Dairy Research Institute, Karnal, Haryana India*

Abstract

Women contribute almost 50 per cent to the population, but still, we are talking about their empowerment, which means we are somewhere lacking in their development. Being women many of us felt various types of differentiation from birth to death. Some rituals are also based on gender. We felt that there is a sort of development in the thinking of people regarding educating women, sending them to different cities for study or work, helping them to start their own small business such as cloud kitchen, art & craft, customized work, home bakery, parlour, etc. family support is extremely important in it. There is an inverted u shape curve we felt when we talk women's empowerment during the Vedic period, the condition of women was good, but it deteriorated as time passed. In the medieval period, it was very bad. Later on, with the help of the help of the constitution, laws, awareness, and education, the situation became become social evils are there that are still restricting women's growth. By over, we can empower our women to empower and make this world a better place to live. The basic and most essential step we all as individuals is to make women aware of their strengths strength and encourage them.

Keywords: *Women empowerment, awareness, development, social evil.*

3.1 Introduction

Women are the backbone of the family. God gave her all the power to give birth, raise the children, do all the household chores, and balance the professional work. But we forget that women are also human beings, not machines. They can't work 24x7 they also need rest as well as proper nutrition. Women in ancient India were very powerful in terms of decision-making, political opinion, education, etc, but later on, the situation of women deteriorated. In the middle era, the situation was at its worst: women were being treated badly by the men in their own families. Currently, the situation is turning better. Women are being treated equally now times but it is not the best scenario we can have. There is still much scope in this field where equal wedges and a good environment can be provided to the women.

Women are the epitome of power, love, sacrifice and courage. Women now play a much different and better role in the modern world. Women nowadays are financially independent, self-sufficient and conscious. They have reached enormous success in every field, whether it is athletics, politics or academia. Co-education has allowed women to march with men in all spheres of society. (Haranath *et al.*, 2021)

Once in the lifespan of women, she thought that it would be easier if she was a man. Since childhood, women have been taught about how to be a good servant for their family. The irony of the situation is that her own mother taught her how to be a good servant. She always prepares to sacrifice. From sacrificing her favourite part of food to her brother to sacrificing her dreams and career for the sake of her family or in-laws. The change should start from family at young age where the mother should treat her children equally, raise them equally, don't divide the work based on their gender.

A country cannot progress until and unless the women of the country take the initiative for development works. The role of women in India is a most talked about issue that is passionately debated every day, as ever more women become conscious of the inequalities and biases they suffer from. Though providing care can empower women if they receive the right training, cultural perceptions of what is and is not "suitable" for a woman's employment often limit the range of options available to them. This outlines the place of women in India and the extent to which they can contribute to society. The boy, however, is free to make any decisions he wants.

When it comes to issue solving, women offer an alternative perspective that improves the quality of the solution. Women contribute a unique perspective that improves the decision-making process when making decisions. I can argue the same point for variety in all its forms. Humanities and sciences faculty will make curriculum decisions better than scientists or humanists working alone. (Trauth, 2020)

In Hindu culture, goddesses has been worshipped equally to god. There are 9 days called Navratri in which Hindus worship goddess, but at the same time, some people who worship the goddess treat the females of their surroundings badly. On the 9th day they worship girls by considering them as a form of goddess but they kill their own female child in the womb of her mother. The increasing rate of female foeticide leads to gender ratio disturbance. afterwards, the government banned checking the gender in the womb. It is a criminal punishable offence nowadays.

The journey from tradition to transformation is a historical retrospective and a forward-looking examination of the contemporary landscape. It delves into the aspirations of modern women, the changing dynamics of gender roles, and the ongoing efforts to dismantle barriers that hinder the full realization of women's potential. Through this exploration, "From Tradition to Transformation" invites readers to reflect on the progress made, the challenges that persist, and the collective vision for a future where women's roles and aspirations continue to evolve, breaking new ground in the ongoing pursuit of equality and empowerment.

3.2 Historical Context of Indian Women

There is a very famous quotation, "Yatra Naryasthu Pujyanthe Ramanthe Tatra Devata," which means, "Where women are given due respect, even gods like to reside there."

3.2.1 Status of Women During Ancient India

- In contrast to modern cultures, women were treated with dignity and respect during the Vedic Period in ancient Greece and Rome.
- They essentially participated in administration, education & politics.
- Among the famous women of the Vedic age were Gargi, Lopamudra, Ghosa and Maitreyi.
- They were viewed as the better half, or Ardhangini, of their husbands.
- Widows were also socially acceptable to remarry.
- In the later Vedic era, the Devdasi practice of women being psycho-sexually committed to the gods became commonplace.
- Women in war were granted more responsibilities and places as well as sexual liberty.
- The Mahabarata and the Ramayana present a vision of kshatriya women fighting in battles to defend their culture; later on, they are also forbidden from entering politics.
- A variety of issues arose during Manusmrithi (the age of Dharmashasthra) as a result of the imposition of numerous barriers to education for women.

- During this time, child marriage held a prominent position in society, as spouses were regarded as gods.
- Remarrying after becoming a widow was strictly deemed illegitimate, and Sati Sahagaman's practice quickly gained popularity.

3.2.2 Status of Women During Medieval Period

- The status of women progressively deteriorated once the Muslim monarchs invaded the nation.
- The purdah system and polygamy were prevalent during this time.
- The purdah system at the time restricted the freedom of women. Their schooling was so impacted by the purdah system.
- The number of child widows increased quickly as a result of the practice of child marriage.
- The community denigrated widows who did not practice "Sati", and the practice itself was promoted.
- For women from lower castes, the situation is considerably worse; they are not even permitted to cover their upper bodies, and if they do, they must pay a fee known as lagaan.

3.2.3 Status of Women in Modern India

- A number of activists made efforts to improve women's status.
- Raja Rammohan Roy's efforts abolished the Sati tradition, which also gave women newfound hope and positive empowerment.
- Ishwar Chandra Vidyasagar's campaign to improve widows' lives resulted in the Widow Re-marriage Act.
- Dr. B. R. Ambedkar's efforts are very much appreciable he gave right to women by giving them legal rights in the constitution.
- They were once dependent on their fathers or husbands, but now that they have forged their own identities, they are self-sufficient.
- Compared to former times, women nowadays are more sensible and realistic.
- Indian women nowadays are more self-reliant and vocal than they have ever been.
- Today's women see themselves as their husbands' real "ardhangini."
- She is more aware of his current environment and the demands of his job.

- Women: in the role of mother, they go from being carers and nurturers to friends.
- Many people believe that having more money gives them the ability to express their views on important choices.
- Women now a days no longer believe that pursuing a profession means sacrificing their family and kids.
- The Indian women is also investing a lot more money in her look.

3.3 Social Evils Practices Performed from Ancient Period till Now

i. **Pardah:** Pardah is a custom first practiced by Muslims and then embraced by many Hindus, particularly in India. It entails keeping women hidden from public view by covering them, including with veils, and using high-walled enclosures, screens, and curtains inside the home.

ii. **Sati:** In the old custom known as sati, a widow would sit atop her husband's funeral pyre as a self-sacrifice. It is debatable if early Hinduism's scriptural mention of it exists, but throughout the Indo-Aryan-speaking regions of India, it has been connected to related Hindu activities.

iii. **Polygamy:** Typically, polygyny results from a man marrying more than one woman;this is known as polygamy. Consequently, polygamy was outlawed in India in 1956 for all Indian residents with the exception of Muslims, who are allowed to have four wives and Hindus residing in Goa and other areas along the western coast, where bigamy is accepted. Hindu marriages that involve polygamy are void.

iv. **Devadasi:** In India, a devadasi is a female artist who is dedicated to the worship and service of a deity or a temple for the rest of her life. The dedication takes place in a ceremony similar to a marriage ceremony.

v. **Child marriage:** One of the violations of human rights is child marriage even though it's illegal, the practice is still common: Before turning 18, one in five girls worldwide is either legally wed or in an unofficial relationship.

vi. **Dowry:** Dowry, which might take the form of gifts or money given by the bridegroom's family to the bride's family after marriage, was a common custom in modern-day India. The prevalence of dowries varies according to class and geography. The bride's family may face severe financial hardship as a result of the dowry system. Because the bride's family was unable to meet their expectations, the in-laws frequently assaulted or killed the bride.

vii. **Female foeticide:** Female foeticide: In India, a most heinous kind of female homicide occurs frequently—even before the woman has a chance to give birth. In India, selective abortion of female foetuses, or female foeticide, results in the death of over a million women per year with devastating effects. This leads to the unavailability of females for weddings in a particular area, so they have to bring brides from other places. Some people kill their newborn baby girl by various methods, such as drowning her in a milk container, smashing her head with a brick, burying her in a ditch, etc. But there are some communities who celebrate the birth of a girl child. They also start a practice in which the plant some trees on the occasion of the birth of girl child.

viii. **Female illiteracy rate:** UNESCO estimates that 129 million girls worldwide do not attend school, comprising 32 million elementary school age students, 30 million lower secondary school students, and 67 million upper secondary school students. Girls who live in conflict-affected nations are more than twice as likely to miss school as girls who live in non-affected countries. Only 49% of the world's various countries have achieved gender parity in basic education. At the secondary school level, when just 24% of nations have attained gender parity in upper secondary education and 42% in lower school, the gender gap widens. India has had a high percentage of female literacy since 2010. The percentage of female literacy in 2010 was 80.35%. Eventually, the rate has gone up with time. India's female literacy rate rose by 14.4% between 2010 and 2021. The rate was 91.95% in 2021. In 2021, there was a 0.6% annual rise in the literacy rate. (Anonymous, 2022)

ix. **Gender discrimination:** Inequalities between men and women in India's health, education, economy and politics are collectively referred to as gender inequality. India is ranked differently on each of these individual categories and on the basis of a number of internationally recognized gender disparity indices, many of which are contentious. India's sex ratio, women's health over the course of their lives, women educational achievement, and even the country's economic situation are all impacted by gender inequality and its social roots. It also keeps laws against rape from being equally applied to men.

x. **Social stigma:** The rejection of or prejudice towards an individual based on observable social traits that set them apart from other members of a community is known as societal stigma.

xi. **Harassment at work place:** The right to equality and dignity of women is flagrantly violated by sexual harassment. Its foundations lie in patriarchy and the belief that men are superior to women, as well as the acceptance of certain types of violence against them. One of these is sexual harassment at work, which regards the various types of harassment as unimportant and harmless. It is frequently justified as "normal" masculine behaviour or "harmless flirtation" those women find enjoyable.

3.4 Solutions to Eradicate the Different Social Evils

i. **Government Policies Favouring Women's Empowerment:** Government policies favoring women's empowerment are pivotal in dismantling systemic barriers and promoting gender equality. These initiatives encompass a range of measures to address historical disparities and foster an environment where women can thrive. Educational policies ensure equal access to schooling and vocational training, empowering women with knowledge and skills. Employment-related measures focus on breaking gender stereotypes, encourage and diversity in the workplace through quotas and incentives. Financial inclusion policies give women access to credit and resources, facilitating entrepreneurship and economic independence. Health initiatives prioritize women's well-being, ensuring access to quality healthcare and family planning services. Legal reforms protect women's rights, addressing issues like gender-based violence and discrimination. Political representation policies aim to boost women's participation in decision-making bodies, fostering a more inclusive governance structure. Social welfare programs, such as maternity benefits and childcare support, provide a safety net for women facing economic challenges. Overall, these government policies collectively contribute to creating a more equitable and empowered society.

ii. **Promoting Gender Equality:** Promoting gender equality is a crucial societal endeavor that seeks to dismantle ingrained stereotypes and foster a world where individuals, regardless of gender, have equal opportunities and rights. Initiatives promoting gender equality encompass various spheres, including education, employment, healthcare, and legal frameworks. Efforts in education aim to eradicate gender-based disparities, ensuring that both girls and boys have equal access to learning opportunities. Workplace policies encourage diversity and inclusivity, challenging gender norms and creating environments where merit takes precedence over gender. Healthcare initiatives address women's specific needs, providing access to reproductive healthcare and promoting overall well-being. Legal reforms target discriminatory practices and enact laws to safeguard women's rights, fostering a more just society. Additionally, awareness campaigns and cultural shifts work to challenge prevailing gender norms, fostering a mindset that values equality. Promoting gender equality is a moral imperative and a strategic investment in creating a more just, prosperous, and harmonious global community.

iii. **Social Awareness:** Social awareness towards women equality is a transformative force that challenges deep-rooted prejudices and fosters a more inclusive society. It involves cultivating a collective consciousness that recognizes the intrinsic value of every individual, irrespective of gender. Initiatives promoting social awareness focus on dismantling stereotypes, debunking myths and fostering empathy to bridge the gender gap. We can break down barriers that perpetuate discrimination by engaging communities in conversations about the importance of women's rights and equality. Social awareness campaigns bring attention to systemic issues such as gender-based violence, unequal opportunities and biased cultural norms, encouraging individuals to question and change these norms. As awareness spreads, a cultural shift occurs, leading to increased support for policies and practices that promote women's empowerment, representation and equal opportunities. Ultimately, social awareness towards women equality is crucial to building a fair and harmonious society where everyone can thrive.

iv. **Promoting Education and Health in Society:** Education is a powerful tool for dismantling gender-based disparities, providing both girls and boys with equal access to knowledge and opportunities. By ensuring that educational systems are inclusive and free from gender bias, we empower women to break through traditional barriers and contribute meaningfully to various fields. Simultaneously, prioritizing health initiatives, including reproductive healthcare, contributes to women's well-being. When women have access to quality healthcare, they can make informed decisions about their bodies and reproductive choices, fostering autonomy. Combining education and health initiatives creates a synergistic effect, as educated women are more likely to prioritize their health and well-being. Together, these efforts empower women individually and promote a more equitable and prosperous society by tapping into the full potential of all its members.

v. **Ensuring fast justice mechanism:** Ensuring a swift and efficient justice mechanism for women in India is paramount to address and deter gender-based crimes and discrimination effectively. Expedited legal proceedings are essential for timely justice, fostering a sense of security among women and reinforcing faith in the legal system. Establishing specialized courts or fast-track mechanisms dedicated to handling cases of crimes against women can help streamline the judicial process. Additionally, enhancing the capacity of law enforcement agencies, promoting sensitivity training and implementing technological advancements can contribute to quicker investigations and trials. An accelerated justice system not only serves as a deterrent for potential offenders but also sends a strong message that crimes against women will be met with swift and severe consequences. This approach not only upholds the principles of justice but also reinforces the commitment to creating a safer and more equitable society for women in India.

3.5 Aspiration of Women from Traditional to Modern

The aspirations of women have undergone a profound evolution from traditional to modern times. In traditional societies, women were often confined to prescribed roles, primarily focused on family and domestic responsibilities. In many cases, the aspiration was to fulfil societal expectations and adhere to established norms.

As societies transitioned to more modern contexts, women's aspirations began to broaden. Education became a gateway to new possibilities, allowing women to aspire to careers, leadership roles, and intellectual pursuits. The desire for financial independence, self-expression, and personal fulfilment gained prominence.

Modern women aspire to equality, inclusivity, and the dismantling of gender stereotypes. Their aspirations now encompass breaking through glass ceilings, challenging societal norms, and contributing to diverse fields. Pursuing meaningful careers, active participation in social and political spheres, and the desire for balanced personal and professional lives are key aspects of modern women's aspirations.

Moreover, contemporary women aspire to inspire future generations, fostering a world where opportunities are not limited by gender. The aspiration is not merely individual but collective striving for a society that recognizes and values the potential of every woman, irrespective of tradition or stereotypes. In essence, the transition from traditional to modern aspirations reflects a journey towards empowerment, choice and realising one's full potential.

3.6 Conclusion

"From Tradition to Transformation: Women's Role and Aspiration" unfolds as a narrative that traverses the historical and cultural landscape, illuminating the dynamic evolution of women's roles and aspirations. This exploration has provided a multifaceted understanding of the challenges, triumphs, and the ongoing journey toward gender equality.

Reflecting on the historical context, it becomes evident that tradition once tightly bound women within defined roles. Yet, through the lens of transformative moments and courageous individuals, we witness the breaking of shackles and the emergence of women who defied norms to reshape their destinies.

The cultural perspectives unveiled in this journey showcase the diversity of women's experiences, underscoring the need to appreciate the intersectionality of gender with other facets of identity. Challenges faced by women, both historically and in the present, highlight the importance of continued efforts to dismantle barriers, confront stereotypes and foster inclusivity.

The triumphs and achievements of women stand as beacons of inspiration, demonstrating the immense potential within every woman when allowed to thrive. The contemporary landscape reflects the aspirations of modern women reaching beyond traditional boundaries, encompassing education, careers, leadership and a quest for a holistic life.

Looking forward, the collective vision is one of equality, empowerment and a world where women's roles and aspirations know no confines. The journey from tradition to transformation is an ongoing narrative and it is the responsibility of society to nurture an environment where women can continue to redefine and fulfil their aspirations. As we celebrate the progress made, the persist challenges remind us of the work yet to be done to ensure a future where every woman can embrace her aspirations with unfettered determination. "From Tradition to Transformation" stands as a testament to women's resilience, strength, and unwavering spirit throughout history, inspiring us to envision a world where their roles and aspirations continue to flourish and shape the tapestry of human progress.

References

Anonymous(2022).https://www.globaldata.com/data-insights/macroeconomic/female-literacy-rate-in india/#:~:text=Female%20Literacy%20Rate%20in%20India%20(2010-2021%2C%20%25),20.00%25%2040.00%25%2060.00&text=In%202010%2C%20 the%20female%20literacy,%2C%20the%20rate%20was%2091.95%25.

Chaudhuri, M. (2012). Indian "Modernity" and "Tradition": A Gender Analysis. Sociological Review. 2(178):281-293.

Das, R. (2020). Women from Ancient to Modern India: An Analysis. IJARIIE. 6(2):1729-1731.

Haranath, B., Dhawal, K., Singh, A. and Parihar, P. (2021). Changing Role of Women from Traditional to Modernity.

Kosec, K., Akramov, K., Mirkasimov, B., Song, J. and Zha, H. (2021). Aspirations and Women's Empowerment: Evidence from Kyrgyzstan. Economics of Transition and Institutional Change. 30(1):101-134.

Krishan, K. (2019). Changing Role of Women in Indian Society. Humanity Science & English Language. 7(36): 9561-9565.

Roy, B. (2022). Women's Empowerment in India, from Ancient Period to Modern Time Period. https://timesofindia.indiatimes.com/readersblog/scatteredthoughts/womens-empowerment-in-india-from-ancient-period-to-modern-time-period-46689/

Sara, N. (2015). Process of Social Transformation among Women in India. Journal ofSocial Sciences Research. 8(2):1586-1590.

Trauth, D. (2020). The Changing Role of Women. https://www.president.txst.edu/denisetrauth/communications/speeches-and-presentations/changing-role-of-women.html.

Whiting, B. B. (2010). Woman's Role in Social Change. Cultural Survival. https://www.culturalsurvival.org/publications/cultural-survival-quarterly/womans-role-social-change.

4

Transforming Lives Measuring the Impact of Technology on Rural Women

Arunima Dhar

Department of Social Work, Rabindra Bharati University, Centre for Distance and Online Education, Kolkata, West Bengal, India

Abstract

The empowerment of rural women is contingent upon narrowing the gender gap and bolstering their participation in the rural economy. This article delves into the crucial role of technology in this pursuit. By facilitating knowledge enhancement and integrating rural women into the global economy, technology has become an indispensable tool for their empowerment. However, to harness its full potential, it is imperative to comprehend the nuanced impact of technology on rural women's lives. This article emphasizes the need to recognize the diverse circumstances of rural women and analyze their experiences with technology comprehensively. By exploring technology's influence on various facets, such as agriculture, entrepreneurship, and daily lives, a holistic understanding of its implications emerges. Nonetheless, entrenched social barriers pose formidable challenges to realizing the benefits of economic independence. Despite advancements, traditional gender norms, and societal expectations continue to constrain women's autonomy and impede their access to education and decision-making. The article underscores the urgency of addressing these social barriers to ensure that economic empowerment translates into substantive empowerment for rural women in India. By elucidating the intersection of technology and gender dynamics in rural settings, this study aims to inform targeted interventions and policy initiatives geared towards fostering inclusive development and empowering rural women.

Keywords: *Gender Gap, Rural Women, Empowerment, Technology, Economic Independence, Social Barriers, Rural Economy.*

4.1 Indian Rural Women Embracing the Changes

Rural areas are chiefly characterized by a strong reliance on primary industries such as agriculture, forestry, fishing, and mining (MacTavish, 2020). When examined by occupation, the agricultural sector frequently acts as the foundation of rural economies, employing a sizable percentage of workers in farming, livestock raising, and crop production (Cagliarini & Rush, 2011). Unlike urban areas, rural regions offer fewer non-agricultural employment opportunities, leading to a greater emphasis on self-employment, small-scale enterprises, and seasonal work patterns. Limited access to formal employment prompts reliance on informal sectors such as casual labour and subsistence farming. Additionally, rural communities experience the outmigration of young individuals seeking better job prospects in urban centers, contributing to demographic shifts and challenges in sustaining rural economies (Philip & Williams, 2019). Further, the gender divides in household activities, which refers to the unequal distribution of domestic responsibilities between men and women within households, reflects entrenched gender roles and societal expectations that persist despite advancements in gender equality. Traditionally, household tasks, such as cooking, cleaning, childcare, and eldercare, have been primarily assigned to women, while men have often been expected to focus on paid employment outside the home. As a result, women often bear a disproportionate burden of unpaid care work, which limits their opportunities for education, employment, and personal fulfillment. At present, traditional gender roles in rural areas are evolving as women increasingly participate in agricultural activities and pursue non-traditional roles, facilitated by factors such as access to education, technology, and changing societal attitudes towards gender equality(Goel, 2023). Despite progress, challenges persist, including limited access to resources and entrenched social norms, necessitating continued efforts to promote women's empowerment and foster inclusive rural development. As a result of these changes, there is a gradual breakdown of traditional gender stereotypes, increased economic empowerment and independence among rural women, improved household resilience, and greater social inclusion and cohesion within rural communities. Increased women's participation in various sectors fosters economic growth, poverty reduction, and gender equality in rural areas, while also contributing to improved food security, community development, and access to health and education services for families and communities. A decreased gender gap and increased participation of women in the rural economy have contributed significantly to the empowerment of rural women(ILO, 2019). Apparently, through active participation in agricultural activities, entrepreneurship, and community development initiatives, rural women have been gaining economic independence, better decision-making power, and enhanced social status within their communities. Moreover, reduced gender disparities in access to education, healthcare, and resources further empower rural women, enabling them to contribute more effectively to their households and communities' well-being (Kabeer, 2005). Overall, these trends lead to greater gender equality, economic resilience, and social inclusion in rural areas, fostering sustainable development and improving rural women's and their families' quality of life.

4.2 Impact of Technology on Rural Women's Lives

The role of technology in our lives is to facilitate communication, streamline tasks, access information, and drive innovation, ultimately enhancing convenience, efficiency, and connectivity in various aspects of daily life. Technology is pivotal in empowering rural women by providing access to information, education, and economic opportunities. Through mobile devices and internet connectivity, rural women can access educational resources, online training programs (Patel & Sayyed, 2014), and market information. This enables them to acquire new skills, enhance their knowledge, and engage in entrepreneurship from remote locations. Additionally, technology facilitates access to agricultural innovation, healthcare services, and advocacy platforms, empowering rural women to improve farming practices, access healthcare remotely, and advocate for their rights and interests. By leveraging technology, rural women also have overcome geographical barriers, enhanced their livelihoods, and participated more actively in economic, social, and political spheres, contributing to inclusive and sustainable development in rural communities (Ahmed, 1983).

This article underscores the significance of exploring technology's impact on rural women's lives. Technology has provided avenues for knowledge enhancement and has contributed to the globalization process, with rural women experiencing its effects as well. Recognizing and analysing the diverse situations would provide a comprehensive understanding of the current scenario and inform efforts to empower rural women through technology.

Fig. 4.1: Empowered by the Lens: Women Farmers and the Tech-Driven Harvest

4.3 Urban and Suburban Influence on Rural Women in India: Shaping Opportunities and Challenges

It has been evidenced that technology can have a transformative impact on women from rural areas and those from rural areas with suburban or urban proximity. While women in rural areas may face limited access to technology and internet connectivity, those in suburban or urban-proximate areas may have better access to these resources. Consequently, rural women with suburban or urban proximity may benefit from a wider range of economic opportunities, access to information, and social connectivity facilitated by technology(Bhatt et al., 2020). However, both groups of women can utilize technology to enhance their access to education, healthcare, and economic opportunities, as well as to connect with peers, access support networks, and advocate for their rights. Bridging the digital divide and promoting digital inclusion is essential to ensure that all women, regardless of their geographic location, can fully leverage the potential of technology for empowerment and socioeconomic advancement (Draude, 2024).

4.4 Empowering Rural Women through Agricultural Technology

Technology has indeed revolutionized the agricultural sector for rural women, viz-a-viz, rural agriculture-based population offering many opportunities for empowerment and economic advancement. Considering rural women in a farming community who rely on traditional farming methods passed down through generations. With access to smartphones and internet connectivity, women can now access crucial information such as weather forecasts, market prices, and agricultural best practices. Armed with this knowledge, they can make informed decisions about crop selection, timing of planting and harvesting, and market sales, ultimately increasing her productivity and income (Watkins et al., 2012). For example, she can strategically plan her farming activities to maximize yields and profits by monitoring weather patterns and market trends through agricultural applications or online platforms. Moreover, technological advancements such as mechanized farming equipment, irrigation systems, and precision agriculture techniques further enhance her productivity and efficiency on the farm. Access to modern farming tools and techniques can streamline their work processes, reduce labour costs, and optimize resource utilization, leading to higher yields and greater economic stability. For instance, she can precisely manage water and fertilizer applications using automated irrigation systems or GPS-guided tractors, minimizing waste and maximizing crop yields (Matholiya et al., 2022). Furthermore, technology empowers rural women to become entrepreneurs in the agricultural value chain. With the rise of e-commerce platforms and online marketplaces, she can directly market her agricultural products to consumers, bypassing intermediaries and capturing a larger share of the market value (Haji, 2021). Additionally, she can explore opportunities for value-added processing businesses, such as making jams, pickles, or handicrafts from her farm produce and selling them online or through local markets. By leveraging mobile banking and microloans, they can access financial services to invest in her business, expand their operations, and improve their livelihood (Gaur & Potnis, 2022). Due to tradition, rural women often work alongside men in farming, contributing to various agricultural tasks. In contrast, their involvement in small and cottage industries is more pronounced, driven by economic necessity and the desire for independence. Despite the transformative potential of technology, challenges such as limited access to internet connectivity, digital literacy barriers, and affordability constraints persist for rural women. By ensuring that all rural women can fully participate in and benefit from the digital agricultural revolution, technology is a powerful tool for empowering agriculture-based women, enhancing productivity, generating income, and improving livelihoods while contributing to sustainable rural development and food security.

Case: Empowering Rural Women Farmers through Mobile Technology in India

In a rural village in India, a mobile technology initiative called "Farm Connect" has revolutionized the lives of women farmers. Developed by a local NGO in partnership with telecommunications companies and agricultural experts, the Farm Connect app provides crucial information such as weather forecasts, market prices, and financial services tailored to the needs of rural women. Through the app, women farmers gain access to real-time agricultural insights, enabling them to make informed decisions about crop management, market sales, and financial investments. As a result, they experience increased agricultural productivity, improved economic stability, and enhanced empowerment. The initiative fosters financial inclusion and builds a supportive community among women farmers, facilitating knowledge sharing and mutual learning. Overall, the Farm Connect initiative exemplifies the transformative impact of technology in empowering rural women farmers and driving sustainable rural development.

4.5 Fostering Rural Women's Entrepreneurship through Technology

Technology has revolutionized rural women's entrepreneurship landscape, offering them opportunities for previously inaccessible growth and expansion. For instance, consider rural women in a remote village who are skilled in traditional handicrafts. With the advent of e-commerce platforms like Etsy or social media platforms like Instagram or Facebook, she can now showcase her products to customers worldwide, breaking free from the confines of her local market and tapping into a global audience. This expanded reach increases her sales potential and introduces her to new trends and customer preferences, enabling her to adapt and innovate her product offerings. Moreover, technology simplifies the day-to-day operations of her business (Ithurbide & Singh, 2022). With the help of mobile apps for inventory management and bookkeeping, women can efficiently track their stock levels, monitor sales, and manage finances, saving time and reducing the burden of administrative tasks (Buteau, 2021). This allows them to focus more on creative aspects of her business, such as designing new products or refining her marketing strategies, driving innovation and differentiation in her market niche. Additionally, technology provides invaluable resources and support for her entrepreneurial journey. Online courses, webinars, and digital libraries offer training on various aspects of entrepreneurship, from marketing strategies to financial management, empowering her with knowledge and skills to run her business effectively (Chaudhuri, Ghosh, & Abichandani, 2020). Furthermore, digital networking platforms enable her to connect with other women entrepreneurs, mentors, and potential collaborators, facilitating knowledge sharing, mentorship, and partnership opportunities. Through online forums and social media groups, she can seek advice, share experiences, and forge valuable connections within a supportive ecosystem of women-led businesses (Rai & Shahila, 2013). Innovation and creativity are further encouraged through technology, with tools such as digital design software and 3D printing enabling her to experiment with new product ideas and bring them to life. Online crowdfunding platforms provide avenues for financing her ventures, allowing her to test the market demand for her innovative products or services before fully launching them (Archana, Kumar M, & Shyamasundar, 2022). In essence, technology serves as a catalyst for empowering rural women entrepreneurs, offering them opportunities for market access, operational efficiency, access to information and resources, networking and collaboration, and fostering innovation and creativity. By embracing technology, rural women entrepreneurs can overcome geographical barriers, expand their businesses, and significantly drive economic growth and sustainable development in their communities.

Case: Empowering Rural Women Entrepreneurs through Gram Shakti

In rural India, the advent of digital platforms has revolutionized entrepreneurship for women, as illustrated by initiatives like "Gram Shakti." Developed by a local NGO and technology firm partnership, Gram Shakti provides rural women with an online marketplace to sell locally sourced products, ranging from handcrafted textiles to organic produce. Through this platform, women entrepreneurs gain access to a wider customer base, transcending geographical constraints and traditional market limitations. Moreover, Gram Shakti offers training programs in digital literacy, financial management, and marketing strategies, empowering women with the skills and knowledge needed to run successful businesses. This initiative fosters economic independence and growth and promotes gender equality and community development in rural India.

4.6 Enhancing Rural Women's Daily Lives through Technology

A rural woman living in a remote village can use her smartphone to access online resources such as educational videos or agricultural extension services (Thakur & Chander, 2018). For instance, she may watch tutorials on sustainable farming practices to improve crop yields or access healthcare information to better care for her family. Using messaging applications like WhatsApp or social media platforms like Facebook can be used by rural women to stay connected with their family members who may live in different parts of the country or abroad. She can share updates, photos, and videos through these platforms, fostering strong social ties despite geographical distance (Mustaqeem, 2020). With entrepreneurial aspirations, rural women can start small businesses selling handmade crafts or agricultural produce online. Using e-commerce platforms like Etsy or local online marketplaces, she can reach customers beyond her immediate community, expanding her market reach and increasing her income. Use a telemedicine platform without needing to travel long distances to a healthcare facility through a mobile health app, rural women can receive medical advice, prescriptions, and follow-up care from the comfort of their homes, saving time and resources (Bhowmik et al., 2013). One interested in learning new skills can enroll in online courses or vocational training programs offered through platforms (Nedungadi et al., 2018) like Coursera or Udemy. For example, she may take a course on digital marketing to enhance her entrepreneurial ventures or learn computer skills to access online job opportunities. Ultimately, rural women managing multiple responsibilities can use productivity apps like Trello or Google Calendar to organize her tasks, set reminders, and prioritize her activities. Using these tools, she can effectively manage her time, balancing household chores, childcare, and personal pursuits more efficiently. These examples demonstrate how technology empowers rural women by providing access to information, communication channels, economic opportunities, healthcare services, educational resources, and tools for time management and efficiency. By leveraging technology in their daily lives, rural women can overcome geographical barriers, expand their opportunities, and improve their overall well-being and quality of life (Masi & Wac, 2022).

Technology integration in agriculture, entrepreneurship, and daily lives has significantly empowered Indian rural women, catalyzing positive transformations across various spheres of their lives. Technology has emerged as a powerful tool for empowering Indian rural women, enabling them to face social barriers, enhance their productivity and livelihoods, and contribute to rural communities' overall development and prosperity.

Case: Improving Daily Lives of Rural Women through Shakti-Tech

In rural India, introducing technology initiatives like "Shakti-Tech" has significantly transformed women's daily lives. Through this initiative, rural women can access mobile devices with essential applications and services tailored to their needs. They can now access vital information on healthcare, agriculture, education, and government schemes in their local languages, empowering them to make informed decisions for themselves and their families. Furthermore, Shakti-Tech facilitates access to telemedicine platforms, educational resources, and financial services, enabling women to improve their health, acquire new skills, and pursue economic opportunities. As a result, rural women are experiencing increased confidence, better health outcomes, enhanced educational attainment, and improved economic stability, contributing to the overall development and empowerment of rural communities in India.

4.7 Identifying Social Barriers for Rural Women in India

Despite achieving economic independence, rural women in India often face persistent social barriers that impede their full participation and empowerment. Deep-rooted gender norms and stereotypes dictate traditional roles and expectations for women in rural communities. Despite being economically independent, women still face pressure to conform to traditional gender roles as homemakers and caregivers, limiting their ability to pursue their own aspirations and opportunities outside of the household. Moreover, many rural women still have limited decision-making power within their families and communities. Patriarchal structures often prioritize the voices and decisions of men, relegating women to subordinate positions and excluding them from important household and community decisions. Social stigma and discrimination based on gender, caste, and class continue to marginalize rural women, particularly those from marginalized communities. Women face discrimination in accessing resources, services, and opportunities, hindering their ability to fully exercise their economic independence and agency (Saha, 2022). Rural women also lack access to supportive social networks and platforms where they can connect with peers, mentors, and allies. Without adequate support systems, especially by their families, women feel isolated and unable to navigate the challenges they face, further reinforcing social barriers to their empowerment. Additionally, many rural women still have limited access to education and information (Bhandari & Smith, 1997), essential for personal and professional development. Illiteracy and lack of awareness about their rights and opportunities may prevent women from fully realizing their potential and asserting their rights. Ultimately, cultural and religious norms constrain rural women's autonomy and agency. Practices such as child marriage, purdah (seclusion of women), and dowry may perpetuate gender inequality and restrict women's freedom and mobility, even if they are economically independent (Dhanaraj & Mahambare, 2019).

4.8 Conclusion

Economic independence and social barriers represent two distinct yet interconnected aspects of the experience of rural women in India. Economic independence refers to the ability of women to generate income and support themselves financially, often through participation in economic activities such as agriculture, entrepreneurship, or wage labour. This economic empowerment enhances women's autonomy, decision-making power, and overall well-being, allowing them to contribute to household income, invest in their families' future, and assert their agency in various spheres of life (Jayachandran, 2021). However, despite achieving economic independence, rural women in India face significant social barriers that limit their full participation and empowerment. These social barriers encompass a wide range of challenges, including deeply entrenched gender norms and stereotypes, limited decision-making power within households and communities, social stigma and discrimination, lack of supportive social networks, limited access to education and information, and cultural and religious constraints. These social barriers can undermine the potential benefits of economic independence, as they restrict women's ability to fully exercise their autonomy, assert their rights, and access opportunities for personal and professional development. For example, even if women are economically independent, they still face pressure to conform to traditional gender roles and expectations, hindering their ability to pursue education, make decisions about their own lives, or participate in public life. Addressing these social barriers is essential for ensuring economic independence translates into meaningful empowerment for rural women in India (Khan, 2022). This requires comprehensive efforts to challenge gender norms, promote women's rights and agency, provide access to education and information, foster supportive social networks, and create an enabling environment that empowers women to participate in and contribute to their communities and societies fully. By addressing these social barriers, rural women can realize their full potential and contribute to rural India's social, economic, and cultural development.

References

Ahmed, I. (1983). Technology and rural women in the Third World. Int'l Lab. Rev., 122, 493.

Archana, M., Kumar M, V., & Shyamasundar, M. S. (2022). Women Entrepreneurship and Innovation in Higher Education: Opportunities and Challenges in India -A Review. Journal of Engineering Education Transformations, 35(3), 202349-2473.

Bhandari, R., & Smith, F. J. (1997). Rural Women in India: Assessment of Educational Constraints and the Need for New Educational Approaches. Journal ofResearch in Rural Education, 13(3), 183-196.

Bhatt, V., Chandrasekhar, S., & Sharma, A. (2020). Regional Patterns and Determinants of Commuting Between Rural and Urban India. Indian Society of Labour Economics, 63, 1041-1063. doi:https://doi.org/10.1007/s41027-020-00276-9

Bhowmik, D., Duraivel, S., Singh, R. K., & Kumar, S. K. (2013). Telemedicine- An Innovating Healthcare System In India. The Pharma Innovation-Journal, 2(4 Part A), 1.

Buteau, S. (2021). Roadmap for digital technology to foster India's MSME ecosystem—opportunities and challenges. CSIT, 9, 233-244. doi:https://doi.org/10.1007/s40012-021-00345-4

Cagliarini, A. & Rush, A. (2011). Economic Development and Agriculture in India. Bulletin, 15-22.

Chaudhuri, S., Ghosh, R., & Abichandani, Y. (2020). Exploring the Risky Terrain of Entrepreneurship With Support From Developmental Relationships: Narratives From Indian Women Entrepreneurs. Advances in Developing Human Resources, 22(2), 137149. doi:https://doi.org/10.1177/1523422320907044

Dhanaraj, S., & Mahambare, V. (2019). Family structure, education and women's employment in rural India. World Development, 115, 17-29. doi:https://doi.org/10.1016/j.worlddev.2018.11.004

Draude, C. (2024). Digital Transformation Through the Lens of Intersectional Gender Research Challenges and Needs for Action. In H. Werthner, C. Ghezzi, J. Kramer, J. Nida-Rumelin, B. Nuseibeh, E. Prem, & A. Stanger (Eds.), Introduction to Digital Humanism (pp. 83-98). Cham, Switzerland: Springer Nature.

Gaur, A., & Potnis, D. D. (2022). Service innovations in mobile banking for creating value for the poor in developing countries. EJISDC, 88(4), e12209. doi: https://doi.org/10.1002/isd2.12209

Goel, R. (2023). Gender gap in mobility outside home in urban India. Travel Behaviour and Society, 32, 100559. doi:https://doi.org/10.1016/j.tbs.2023.01.004

Haji, K. (2021). E-commerce development in rural and remote areas of BRICS countries. Journal of Integrative Agriculture, 20(4), 979-997. doi:https://doi.org/10.1016/S2095-3119(20)63451-7

ILO. (2019). Empowering Women in the Rural Economy. Genva: International Labour Organization. Retrieved from https://www.ilo.org/wcmsp5/groups/public/-ed_dialogue/-sector/documents/publication/wcms_601071.pdf

Ithurbide, C., & Singh, K. M. (2022). Digital Platforms and Craft Workers in India in the Time of COVID. South Asia Multidisciplinary Academic Journal [Online]. doi:https://doi.org/10.4000/samaj.8198

Jayachandran, S. (2021). Social Norms as a Barrier to Women's Employment in Developing Countries. IMF Economic Review, 69, 576-595. doi:https://doi.org/10.1057/s41308-021-00140-w

Kabeer, N. (2005). Gender Equality and Women's Empowerment: A Critical Analysis of the Third Millennium Development Goal. Gender and Development, 13(1), 13-24.

Khan, S. (2022). Analyzing the career development barriers faced by rural women in India. Problems and Perspectives in Management, 20(1), 143-152. doi:http://dx.doi.org/10.21511/ppm.20(1).2022.13

MacTavish, K. A. (2020). Rural Communities. International Encyclopedia of Human Geography (Second Edition), 9-14. doi:https://doi.org/10.1016/B978-0-08-102295-5.10312-9

Masi, A. D., & Wac, K. (2022). The Importance of Smartphone Connectivity in Quality of Life. In K. Wac, & S. Wulfovich (Eds.), Quantifying Quality of Life (pp. 523-551). Health Informatics. Springer, Cham. doi:https://doi.org/10.1007/978-3-030-94212-0_23

Matholiya, C. S., Gupta, P., Agravat, V. V., Patel, U. V., & Balas, P. R. (2022). Automatic guidance systems in agricultural autonomous robotic machine: A review. The Pharma Innovation Journal, 11(3): 307-312.

Mustaqeem, M. (2020). Working abroad while families at home: Negotiating patriarchal role for women in South Asia: A case study of Gopalganj district of Bihat, India. Journal of South Asian Studies, 8(3), 135-145. doi:https://doi.org/10.33687/jsas.008.03.3745

Nedungadi, P. P., Menon, R., Gutjahr, G., Erickson, L., & Raman, R. (2018). Towards an inclusive digital literacy framework for digital India. Education + Training, 60(6), 516528. doi:https://doi.org/10.1108/ET-03-2018-0061

Patel, S., & Sayyed, I. U. (2014). Impact of Information Technology in Agriculture Sector. International Journal of Food, Agriculture and Veterinary Sciences, 4(2), 17-22.

Philip, L., & Williams, F. (2019). Remote rural home based businesses and digital inequalities: Understanding needs and expectations in a digitally underserved community. Journal of Rural Studies, 68, 306-318. doi:https://doi.org/10.1016/j.jrurstud.2018.09.011

Rai, G. A., & Shahila, Z. (2013). Rural India : The Next Frontier for Social Media Networks. International Journal of Engineering Research & Technology (IJERT), 2(1), 1-6.

Saha, A. K. Social, Cultural and Religious Barriers of Women: How it affects their progress in India. Parikalpana: KIIT Journal of Management(15), 29-44. doi:10.23862/ kiit-parikalpana/2019/v15/i1-2/190172

Thakur, D., & Chander, M. (2018). Use of Social Media in Agricultural Extension: Some evidences from India. International Journal of Science, Environment, 7(4), 1334 - 1346.

Watkins, J., Kitner, K. R., & Mehta, D. (2012). Mobile and smartphone use in urban and rural India. Continuum, 26(5), 685-697. doi:10.1080/10304312.2012.706458

5

Agriculture and Technology Empowering Women Farmers for Sustainable Development

***Shaik Sameer*[1] *and Mutha Sravya*[2]**

[1]*Department of Agronomy, Koneru Lakshmaiah Education Foundation, Guntur Andhra Pradesh, India*

[2]*Department of Agronomy, SR University, Warangal, Telangana, India*

Abstract

This chapter discusses the important role of women in agriculture and the need to empower them for sustainable development. It looks at the historical contributions of women to farming, the challenges they face today, and the initiatives by the Government of India to address gender disparities in agriculture. The study emphasizes the importance of empowering women by giving them agency, access to resources, and involvement in decision-making in the agricultural sector. By addressing gender disparities, societies can unlock the full potential of women farmers for equitable and sustainable agricultural development.

Keywords: *Women empowerment, agriculture, sustainable development, gender equality, empowerment frameworks, gender-sensitive indicators.*

5.1 Introduction

"Women give life and play a vital role in safeguarding the ecosystem. Technology is the Plow that furrows the soil of justice, enabling women farmers to plant the crops of sustainable development."

Women are half the people in the world, and they have done better than men in many ways in the last 30 years. But they still face more problems than men, because they do not have the same access to things like money, land, and power. Gender equality is not only a good thing by itself but also good for the economy, society, and the environment. If women and men have the same chances to do things and make decisions, we can have a better way of development that considers the needs of both women and men.

In India, women are integral to the agricultural sector, making a substantial contribution to both food production and the rural economy. Women still make up 42% of the agricultural workforce, but they face obstacles such as restricted access to resources and decision-making processes and low land ownership (12.8 per cent vs 31.6 per cent for men).

Studies show that many female farmers are unable to access contemporary agricultural technologies, loans, or extension services, which reduces their output. For example, an ICRISAT report emphasizes how little women know about modern farming technology and techniques.

Regarding these obstacles, the Indian government has launched initiatives such as the Mahila Kisan Sashaktikaran Pariyojana (MKSP), emphasizing the enhancement of women's agricultural skills and resource accessibility. Furthermore, programs like Mahila Shakti Kendra (MSK) and Rashtriya Mahila Kisan Diwas seek to assist and increase awareness of rural women's involvement in agriculture.

5.2 History

Agriculture is the practice of growing plants and animals for food and other uses. It is one of the most important inventions in human history, but who started it? Many researchers believe women were the first farmers about 10,000 years ago. Women were usually the ones who gathered wild plants for food, and they learned how to save and plant the seeds of the best ones. They also cared for the animals and domesticated some of them, like goats and sheep. By doing this, women created the first crops and livestock and made farming possible. Farming changed how people lived because they could settle in one place and produce more food than before. Women played a key role in this change and continue to be involved in agriculture today.

5.3 Woman Empowerment

As a complex and evolving process, empowerment enables individuals to take charge of their lives and actively address issues that matter to them. It involves developing skills, confidence, and a sense of agency, allowing people to make informed decisions and impact their personal, social, and professional lives.

In the realm of gender equality, women's empowerment is vital for breaking down traditional norms and systemic barriers. It advocates for a shift where rights, responsibilities, and opportunities are not dictated by gender but are based on principles of equality and fairness. By promoting women's empowerment, societies aim to ensure that individuals, regardless of gender, can shape their destinies, contribute to societal progress, and access resources and opportunities equally.

The concept of gender equality challenges the idea that roles and expectations are fixed by biological differences. Instead, it acknowledges that these roles are socially constructed and can change across cultures and over time. Gender roles and expectations are learned through social interactions, cultural practices, and institutional structures. Recognizing this variability is crucial for promoting inclusivity and understanding that preconceived gender notions should not confine individuals' potential.

As societies progress, the understanding that gender roles are not fixed allows for redefining norms and expectations, creating environments that embrace diversity, and ensuring fair distribution of opportunities and responsibilities. Women's empowerment thus becomes pivotal in challenging and reshaping these learned gender roles, fostering a more inclusive and just society where everyone can fully participate and thrive.

5.4 Sustainable development

Sustainable development is an idea that tries to make sure that people do not use up or damage the things that future people will need. It is based on the understanding that the Earth has limited resources and living things that we need to take care of and use smartly.

Sustainable development has three main aspects: economic, social, and environmental. These are the areas where human actions affect the world and where we need to balance different goals and outcomes. Economic development means making more money and improving the quality of life. Social development means supporting human dignity, fairness, and rights. Environmental development means conserving nature and reducing the effects of global warming.

Some experts have proposed to include cultural diversity as another aspect of sustainable development. This means that culture is not just a way of expressing who we are and what we believe but also finding new solutions and ideas. Cultural diversity also means valuing the different ways of seeing, knowing, and doing that influence how people relate to nature and each other. An integrated and holistic approach to sustainable development means that the three or four pillars are not treated as separate or competing goals but as interrelated and mutually reinforcing dimensions of human well-being. It also means that the impacts and benefits of human actions are assessed across different scales, sectors, and stakeholders and that the trade-offs and synergies are identified and addressed. An integrated and holistic approach requires collaboration, coordination, and participation among different actors and institutions and multidisciplinary and transdisciplinary methods and tools.

Sustainable development is a concept that became famous after the Brundtland Report in 1987, which was written by a group of experts on environment and development, led by the former Prime Minister of Norway, Gro Harlem Brundtland. The Sustainable Development Goals (SDGs) are 17 global goals and 169 targets that the United Nations agreed on in 2015, as part of the 2030 Agenda for Sustainable Development. The SDGs deal with many issues, such as poverty, health, hunger, education, gender equality, energy, climate change, nature, and peace.

5.5 Empowering Women Farmers for Sustainable Development

Women play a crucial role in rural areas, where agriculture and related sectors are vital for livelihoods. They make up 80 percent of the economically active population. Among them, 33 percent work as agricultural labourers, while 48 percent are self-employed farmers actively participating in the agricultural value chain. From pre-harvest production to post-harvest processing, packaging, and marketing, rural women enhance overall agricultural productivity.

According to Pingali *et al.* (2019), the increasing participation of women in the agricultural sector has raised the ratio of women to men and significantly contributed to the per capita GDP. Recognizing the importance of women as a vital demographic for sustainable food systems, the Food and Agriculture Organization (FAO, 2011) projects that implementing women-oriented reforms ensuring equal access to resources, skill development, and opportunities in agriculture could potentially boost agricultural output in developing countries by 2.5 to 4 percent.

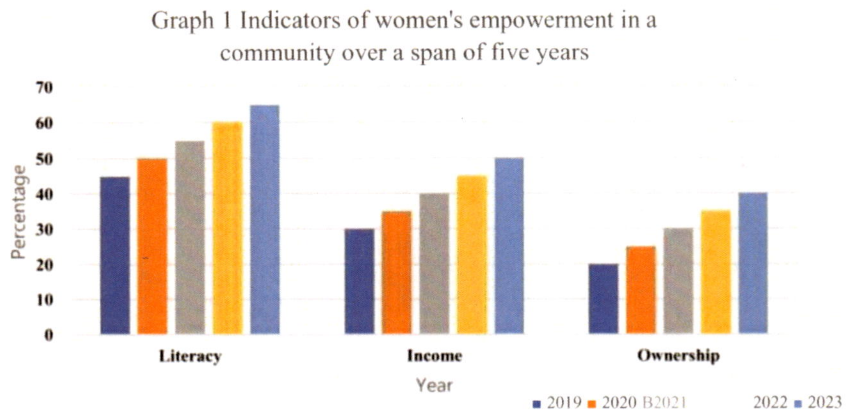

Fig. 5.1: Indicates Women's empowerment in community over a span of 5 years (*Source:* United Nations (UN) Women and World Bank)

Recent announcements under the Aatma Nirbhar Bharat initiative have allocated funds to support women involved in agricultural development, aiming for equal provision of rural services. However, challenges persist for women engaged in farming activities, particularly regarding their limited knowledge of new agricultural tools and technologies. Insufficient awareness hinders their ability to leverage these advancements, necessitating support from extension officers and Krishi Vigyan Kendra (KVK) officials to empower women in agriculture. Research from the U.N. Food and Agriculture Organization emphasizes that providing women with the same access to non-land resources as male farmers could lift 100-150 million people out of poverty. The involvement of women in agriculture in rural India is extensive, with their roles ranging from managers to landless labourers. Their contribution, estimated at 55 to 66 percent of the total labour in farm production, underscores their indispensable role in various regions and agro-production systems.

Despite making up about 30 percent of the farm workforce, women actively participate in daily activities, engaging in physically demanding tasks such as weeding, seedling transplantation, and cereal harvesting. Their daily commitment to agricultural operations ranges from 8 to 9 hours during peak seasons and 4 to 5 hours during slack seasons. Women's involvement spans from land preparation to harvesting, showcasing their versatility in traditional tools and techniques.

In essence, the narrative highlights rural women's dynamic and indispensable role in agriculture, emphasizing the need for comprehensive support, education, and empowerment initiatives to harness their full potential in driving sustainable agricultural development.

According to the World Bank, women account for 42 percent of the agricultural labor force in India, but only 12.8 percent of their own land, compared to 31.6 percent of men. The Indian government has launched several schemes and programs to empower women in agriculture, such as the Mahila Kisan Sashaktikaran Pariyojana (MKSP), which aims to improve women farmers' skills, knowledge, and capacities. The International Crops Research Institute for the Semi-Arid Tropics (ICRISAT) has developed a mobile app called GreenPHABLET, which provides women farmers with information on weather, pest management, crop prices, and government schemes. The app enables women to connect with other farmers and experts through voice and video calls.

5.6 Initiatives by Government of India

i. **Mahila Shakti Kendra (MSK):** The Indian government has launched a new program called Mahila Shakti Kendra from 2017 to 2020 to support rural women through community involvement and help them achieve their full potential. This initiative will give rural women access to government benefits and empower them through training and skill development. In 115 of the most underdeveloped districts, the program aims to engage the community through college student volunteers as part of the MSK Block-level activities. These volunteers will have a major role in spreading awareness about important government programs and social issues. They will also have the choice to join NSS/NCC cadre students. At the block level, the program will offer a chance for student volunteers to contribute to the development process and ensure that women are equal partners in India's progress.

ii. **Mahila Kisan Sashaktikaran Pariyojana (MKSP):** Launched by the Ministry of Rural Development, MKSP focuses on empowering women in agriculture by enhancing their participation in decision-making, skill development, & providing them with better access to resources.

iii. **Rashtriya Mahila Kisan Diwas:** Observed on October 15th, this day recognizes the contribution of women farmers in agriculture. Various awareness programs and events are organized to highlight their role and address the challenges they face.

iv. **Sub-Mission on Agricultural Extension (SMAE) Gender Component:** This initiative under the National Mission on Agricultural Extension & Technology (NMAET) addresses gender-related agricultural issues by promoting gender-sensitive extension services, training, and capacity building.

v. **National Livestock Mission (NLM) - Women Component:** NLM includes a specific component focusing on women's participation in livestock farming. It supports women's self-help groups (SHGs) involved in livestock-related activities.

vi. **National Rural Livelihood Mission (NRLM):** NRLM emphasizes the economic empowerment of rural women by promoting SHGs. It provides financial assistance, skill development, and market linkages to enhance women's participation in various livelihood activities, including agriculture.

vii. **Pradhan Mantri Fasal Bima Yojana (PMFBY):** While not exclusively focused on women, PMFBY is a crop insurance scheme that benefits all farmers, including women. It provides financial protection to farmers in case of crop failure and encourages sustainable agricultural practices.

viii. **Paramparagat Krishi Vikas Yojana (PKVY):** This scheme promotes organic farming and is open to all farmers, including women. It encourages the use of traditional and organic methods for sustainable agriculture.

5.7 Recognizing and addressing gender disparities

i. **Literacy:** This indicator measures the level and quality of education women have attained, such as literacy, numeracy, or vocational skills. It also captures the extent to which women can access and benefit from educational opportunities, such as formal, informal, or lifelong learning. It reflects women's human capital and potential for improving their well-being and empowerment.

ii. **Beyond the Fields:** This phrase implies that the paper goes beyond the conventional measures of women's empowerment in agriculture, which refers to the extent to which women can make choices and act on them regarding agricultural activities, such as what crops to grow, how much to sell, and how to use the inputs. It also captures women's degree of freedom and agency in carrying out their production tasks without being constrained by social norms or expectations.

iii. **Women's control in agriculture:** Women can influence and manage various aspects of agricultural production, such as inputs, outputs, resources, and income. It also captures women's degree of autonomy and agency in pursuing their goals and interests in the agricultural sector.

iv. **Women's empowerment in agriculture:** The expansion of women's ability to make strategic choices and act on them in the agricultural sector. It also captures the degree of agency and influence women have over their own lives and resources, as well as their roles and contributions to society.

v. **Time Equity:** This indicator shows how much and how hard women work in farming and other areas, such as home, care, or community work. It also shows how well women can manage their work and personal time, and have some free time. It reflects women's costs and choices in using their time and energy for different activities.

1.8 Conclusion

Empowering women farmers is morally right and necessary for sustainable agricultural development. Even though women have historically led the way in agriculture, their achievements are sometimes overlooked. Societies may gain from the full potential of women farmers for economic growth, social fairness, and environmental resilience by supporting gender equality in agriculture. Women's rights in agriculture may be advanced through the Government of India's efforts, empowerment frameworks, and gender-sensitive indicators. In order to build an inclusive and sustainable agricultural environment where men and women have equal opportunities to grow and contribute to a successful future for all, it is critical to overcome gender inequities.

References

Deepika Sharma, Rashmi Chaudhary and Krishan Kumar (2020). Women Empowerment in Agriculture and its Dimensions: A Conceptual Review.International Journal of Economic Plants 7(3):111-114.

Gupta, S., Pingali, P.L., Andersen, P.P. (2017). Women's empowerment in Indian agriculture: does market orientation of farming systems matter? Food Security. 9:1447-1463.

Lisa Warth and Malinka Koparanova (2012). Empowering women for sustainable development. United Nations Economic Commission for Europe (UNECE) discussion paper series.

Logeswari, S. and Thiruchenduran, S. (2016). Empowerment of Women Farmers for Agricultural Development. Imperial Journal of Interdisciplinary Research. 2(8):2454-1362.

Majumdar, J., Shah, P. (2017). Mapping the role of women in agriculture. Annals of Anthropological Practice 41: 46-54.

Ministry of Women and Child Development. Mahila Shakti Kendras (MSK) | Ministry of Women & Child Development|IN|sfymu (wcd.nic.in)

Neelam Patel and Tanu Sethi (2021). Rural Women: Key to New India's Agrarian Revolution. Kurukshetra.

Pingali P., Abraham A. A. M., and Rahman A. (2019). Transforming Food Systems for a Rising India. Palgrave Studies in Agricultural Economics and Food Policy. ISBN 978-3-030-14408-1 ISBN 978-3-030-14409-8 (eBook) https://doi.org/10.1007/978-3-030-14409-8.

Role of women in farming in India - INSIGHTSIAS. https://www.insightsonindia.com/social-justice/issues-related-to-women/women-in-agriculture/role-of-women-in-farming-in-india/.

Singh, S., Arora, R. (2010). Ergonomic intervention for preventing musculoskeletal disorders among farm women. Journal of Agricultural Sciences. 1(2): 61-71.

United Nations (UN) Women:https://data.unwomen.org/World Bank: Gender and empowerment. https://data.worldbank.org/

6

Beyond Connectivity Digital Literacy and Skill Development

Nikhil Kumar[1], Amrit Warshini[2] and Aman Verma[2]

[1]Department of Extension Education, Bihar Agricultural University, Sabour Bhagalpur, Bihar, India

[2]Department of Extension Education, Acharya Narendra Deva University of Agriculture and Technology, Kumarganj, Ayodhya, Uttar Pradesh, India

Abstract

Rapid technological growth and digital change have defined the modern period, and digital literacy and skill development have become essential preconditions for individual empowerment and societal advancement. In order to effectively navigate the complexity of the digital world, it is imperative that digital literacy and skill development be explored in all its facets. Digital literacy includes the critical thinking, evaluation, analysis, and creation of digital material and the fundamental skills needed to use digital devices and apps. In a world where communication is becoming more and more integrated, it helps people to access, understand, and convey information efficiently. Furthermore, the development of critical thinking, problem-solving, and creative skills all necessary for success in the digital age is facilitated by digital literacy. Concurrently, innovation and economic competitiveness now depend heavily on developing digital skills. In today's work environment, proficiency in fields like cybersecurity, digital marketing, data analysis, and coding is highly valued. People with these abilities will be in a better position to adjust to the changing needs of the digital economy and make significant contributions to technical progress.

Keywords: *Digital Literacy, Skill, ICT, Digital Platforms, Digital Awareness.*

6.1 Introduction

In its broadest sense, literacy stands as a fundamental pillar of youth development, transcending mere reading and writing abilities to encompass critical skills necessary for navigating today's complex world (UNESCO, 2020). Concurrently, connectivity serves as the backbone of digital transformation, facilitating communication, collaboration, and access to information on a global scale (ITU, 2021). Skill development, particularly in digital literacy, has become indispensable in an increasingly digitized society, empowering individuals to thrive in both personal and professional domains (European Commission, 2019).

i. **Literacy in the Digital Age:** Literacy, traditionally defined as the ability to read and write, now encompasses a spectrum of skills essential for success in the digital era (UNESCO, 2020). This includes the capacity to evaluate information critically, discern fact from fiction, and engage with diverse media forms (UNESCO, 2017). Furthermore, literacy is the gateway to education, empowering youth to excel academically and laying the groundwork for future success (UNESCO, 2020).

ii. **Connectivity and Its Implications:** Connectivity, referring to the ability of devices and systems to establish communication, is pivotal in facilitating digital interactions and access to online resources (ITU, 2021). Digital connectivity has become a basic necessity, akin to water or electricity, particularly in light of the global pandemic, which has underscored the importance of remote access to education, work, and essential services (UNESCO, 2020). However, disparities in access to digital technologies persist, exacerbating existing inequalities, particularly among marginalized populations (ITU, 2021).

iii. **Digital Literacy and Skill Development:** Digital literacy, encompassing a range of literacies such as ICT/computer literacy, information literacy, and media literacy, is essential for navigating the digital landscape effectively (European Commission, 2019). Skill development in digital literacy empowers individuals to harness the potential of digital tools ethically, responsibly, and inclusively (European Commission, 2019). In India, for instance, initiatives such as the Universal Payment Interface (UPI) and Aadhaar highlight the importance of digital literacy in fostering inclusive growth and innovation (KPMG, 2020).

6.2 Definition of Digital Literacy

- It is defined as "the ability to understand and use information in multiple formats from a wide range of sources when presented via computers."
- Digital literacy is accessing, evaluating, and utilizing digital technologies effectively and responsibly. It encompasses many competencies, including technical proficiency, information literacy, media literacy, cybersecurity awareness, and critical thinking.
- Digital literacy is the capacity of people and communities to comprehend and apply digital technology for worthwhile purposes in everyday life.

Simply put, it's the capacity to use a computer, smartphone, or the internet for daily tasks and to establish online connections with other people.

The goal of digital literacy is to provide trainees with fundamental ICT skills that are pertinent to their requirements. This will allow them to use IT and associated applications to actively engage in the democratic process and further improve their employment chances.

6.3 Various Ways for Internet Facility

- **Cable:** Cable TV providers can transmit the internet over their existing physical connections.
- **Digital subscriber line (DSL):** DSL can provide a similar level of service as cable but without a cable subscription.
- **Mobile handset:** Connect to the internet via a mobile handset.
- **Wi-fi hotspots:** Connect to the internet via wi-fi hotspots.
- **Low-orbiting satellite signals:** Connect to the internet via low-orbiting satellite signals.

Fig. 5.1: Figure shows the various ways for digital learning.
(*Source:* https://www.eppingforestdc.gov.uk/community/digital-skills-learning-support/)

6.4 Aspects of Digital Literacy

i. **The Power of Internet:** Internet usage is a gateway to global connectivity and knowledge dissemination (Smith, 2020). Responsible and constructive internet usage offers myriad advantages, fostering societal relationships and strengthening familial bonds (Jones, 2019). Social media platforms facilitate connections among individuals, including rekindling old friendships or professional relationships (Adams, 2018). The internet's vast potential enables e-governance, offering inclusive access to governmental services and resources (Johnson, 2021).

ii. **Use of Emails:** Email has revolutionized communication, rendering traditional means such as telegrams obsolete (Brown, 2017). Its speed, reliability, and simplicity make it an indispensable tool for dialogue and information exchange (Clark, 2019). Establishing an email account facilitates swift communication and access to government programs for employees (Taylor, 2020).

iii. **Uses of Apps:** Computer applications, or apps, streamline various tasks, enhancing user experience (White, 2018). These user-friendly programs cater to specific functions, simplifying complex operations (Roberts, 2019). The widespread adoption of apps has transformed daily life, offering convenience and efficiency to users (Miller, 2020).

iv. **Uses of Browser:** Browsers are the primary gateway to online resource access (Garcia, 2018). Different browsers offer unique features to accommodate diverse user preferences (Lee, 2019). Users must select appropriate browsers based on their needs and requirements (Patel, 2021).

v. **Uses of Government Applications:** The government's emphasis on digital governance has led to the development of operation-specific applications like Umang and Yono (Hernandez, 2020). These user-friendly platforms facilitate access to government schemes and programs, promoting financial inclusion and transparency (Sanchez, 2021).

vi. **Uses of Unified Payment Interface (BHIM):** The Unified Payments Interface (UPI) enables seamless transactions through apps like BHIM, simplifying financial transactions (Nguyen, 2018). Users can transfer funds instantly across banks using their mobile phones or Virtual Payment Addresses (UPI ID) (Chen, 2020).

vii. **Ethical Uses of Internet:** Internet users must exercise caution to prevent harm, refraining from sharing content that could incite violence or cause harm (Smith, 2020). The dissemination of inappropriate content, including pornography, necessitates regulatory measures to protect vulnerable individuals (Jones, 2019).

viii. **Threats of Cyber Security:** Cybersecurity threats pose significant risks to individuals and organizations, requiring vigilance and protective measures (Brown, 2017). Safeguarding sensitive information, such as OTPs and passwords, is essential to mitigate cyberattacks (Clark, 2019).

6.5 Use of ICT In Digital Literacy

In today's interconnected world, the utilization of Information and Communication Technology (ICT) has become increasingly prevalent in the realm of digital literacy and skill development. ICT provides robust tools and platforms that can revolutionize how individuals learn, develop, and apply skills essential for personal, academic, and professional growth (Czerniewicz *et al.*, 2019). This essay elucidates the multifaceted ways ICT is employed to foster digital literacy and skill enhancement.

i. Online Learning Platforms represent a significant facet of ICT-enabled digital literacy initiatives. These platforms, including Coursera, Udemy, Khan Academy, and LinkedIn Learning, offer diverse courses and resources tailored to varying interests and skill levels (Ally, 2004). Learners can access interactive tutorials and materials covering an extensive spectrum of subjects, ranging from technical proficiencies such as coding and data analysis to soft skills like communication and leadership (Ally, 2004).

ii. Virtual Classrooms and Webinars facilitated by ICT have emerged as indispensable tools for remote learning and collaboration. Platforms such as Zoom, Google Meet, and Microsoft Teams enable individuals to engage in live lectures, discussions, and interactive sessions from anywhere (Kozinets, 2010). These platforms foster dynamic learning experiences, transcend geographical constraints and promote active participation among learners.

iii. Digital Libraries and Open Educational Resources (OER) are another critical component of ICT-driven digital literacy initiatives. Platforms such as OpenStax, Project Gutenberg, and Google Scholar provide access to an extensive array of educational materials, textbooks, articles, and multimedia resources at minimal or no cost (Weller, 2014). Learners benefit from a wealth of content conducive to self-directed learning and research endeavors.

iv. Interactive Multimedia Content facilitated by ICT enhances engagement and comprehension in educational settings. Platforms like YouTube, TED-Ed, and Kahoot! offer immersive learning experiences through videos, simulations, animations, and gamified content (Mayer, 2014). These multimedia formats cater to diverse learning styles, fostering deeper understanding and knowledge retention.

v. Mobile Learning Applications represent a burgeoning frontier in digital literacy promotion. Apps such as Duolingo, SoloLearn, and Memrise directly deliver educational content and activities to learners' mobile devices, enabling convenient access to language learning, coding tutorials, and other skill-building resources (Kukulska-Hulme & Traxler, 2007). Mobile learning empowers individuals to engage with educational content anytime, anywhere, facilitating continuous learning.

vi. Social Media and Online Communities harnessed by ICT provide collaborative learning and knowledge-sharing avenues. Platforms like Reddit, Quora, and LinkedIn groups serve as virtual forums for networking, mentorship, and professional development (Rheingold, 2012). Learners leverage these online communities to connect with peers and experts, fostering a culture of lifelong learning.

vii. Digital Skills Training Programs equip individuals with the proficiencies necessary to thrive in the digital age. Government initiatives, corporate training programs, and community organizations offer workshops, boot camps, and courses to enhance participants' digital literacy and technological competence (Livingstone & Helsper, 2007). These programs are crucial in bridging the digital divide and promoting digital inclusion.

ICT is a transformative force in digital literacy and skill development, democratizing access to education, fostering collaboration and innovation, and empowering individuals to realize their full potential in an increasingly digitalized society (Wong & Looi, 2011). By harnessing ICT effectively, stakeholders can create inclusive, engaging, and impactful learning experiences that equip learners with the competencies needed to thrive in the 21st century.

6.6 What is Skill development

Skill development refers to the process of acquiring, enhancing, and refining the abilities, competencies, and expertise necessary to perform specific tasks or roles effectively. It encompasses various capabilities, including technical skills, soft skills, and specialized knowledge, tailored to meet the demands of various industries, professions, and roles.

Skills can be broadly categorized into two main types:

i. Hard Skills: Hard skills are specific, measurable abilities typically acquired through formal education, training, or practical experience. These skills are often technical and can include proficiency in programming, engineering, accounting, healthcare procedures, and mechanical operations.

ii. Soft Skills: Soft skills, also known as interpersonal or people skills, are non-technical skills that enable individuals to interact effectively with others and navigate interpersonal dynamics. Examples of soft skills include communication, teamwork, leadership, problem-solving, adaptability, time management, and emotional intelligence.

Skill development is a continuous process that occurs throughout life. It encompasses formal education, vocational training, on-the-job experiences, self-directed learning, and professional development activities. Skill development is essential for personal growth, career advancement, and adaptability in an ever-changing labor market.

6.6.1 Skill development initiatives can take various forms, including:

i. Formal education programs offered by schools, colleges, and universities.
ii. Vocational training programs designed to equip individuals with job-specific skills and certifications.
iii. Apprenticeships and internships that provide hands-on learning experiences under the guidance of experienced professionals.
iv. Online courses and e-learning platforms that offer flexible, self-paced learning opportunities.
v. Professional development workshops, seminars, and conferences aimed at enhancing existing skills and acquiring new ones.
vi. Government-sponsored skill development programs aimed at addressing unemployment, poverty, and socio-economic disparities.

6.6.2 Digital Instrument for Skill Development

Digital instruments for skill development encompass a wide range of technological tools and platforms designed to facilitate learning, practice, and mastery of various skills. These instruments leverage digital technologies to offer interactive, engaging, and personalized learning experiences tailored to learners' needs and preferences. Here are several examples of digital instruments for skill development across different domains:

i. **Online Learning Platforms:** Platforms like Coursera, Udemy, and Khan Academy offer a diverse array of online courses, tutorials, and resources covering a wide range of subjects and skills, from technical skills such as programming and data analysis to soft skills such as communication and leadership. Learners can access multimedia content, interactive exercises, and assessments to enhance their knowledge and proficiency in specific areas.
ii. **Mobile Applications:** Mobile apps provide convenient and accessible platforms for skill development on the go. Apps like Duolingo for language learning, SoloLearn for coding, and Lumosity for cognitive skills training offer interactive exercises, quizzes, and games to help users acquire and practice skills anytime, anywhere, using their smartphones or tablets.
iii. **Virtual Reality (VR) and Augmented Reality (AR) Simulations:** VR and AR technologies offer immersive, experiential learning environments that simulate real-world scenarios and enable hands-on

practice of skills in a safe and controlled setting. VR training programs are increasingly used in healthcare, manufacturing, and aviation fields to train professionals and enhance technical skills.

iv. **Gamified Learning Platforms:** Gamification platforms like Kahoot! and Quizizz gamify the learning experience by incorporating game elements such as quizzes, challenges, and rewards to engage learners and motivate them to achieve learning objectives. These platforms make skill development interactive, competitive, and enjoyable, fostering deeper engagement and knowledge retention.

v. **Online Coding Platforms:** Platforms like Codecademy, LeetCode, and HackerRank provide interactive coding challenges, tutorials, and coding exercises to help users learn programming languages, algorithms, and problem-solving techniques. These platforms offer immediate feedback and performance analytics to track progress and identify areas for improvement.

vi. **Digital Collaboration Tools:** Collaboration tools such as Google Workspace, Microsoft Teams, and Slack facilitate teamwork, communication, and project collaboration among individuals and teams. These tools enable users to collaborate on projects, share resources, and coordinate tasks in real-time, enhancing collaboration skills and promoting effective teamwork.

vii. **Simulation Software:** Simulation software applications like SimScale, ANSYS, and MATLAB offer virtual environments for engineering, scientific, and technical simulations. These tools allow users to simulate complex systems, conduct experiments, and analyze data, helping them develop practical skills and problem-solving abilities in their respective fields.

viii. **Digital Art and Design Tools:** Creative software applications like Adobe Creative Cloud, Procreate, and Canva provide digital tools and resources for graphic design, digital art, and multimedia production.

These tools enable users to develop creative skills, explore visual storytelling techniques, and express their ideas through digital media.

6.6.3 Digital Platforms for Skill Development

Digital platforms for skill development have emerged as powerful tools for individuals seeking to enhance their competencies, advance their careers, and adapt to the demands of the modern workforce. These platforms leverage technology to provide flexible, accessible, and personalized learning experiences across a wide range of subjects and disciplines. Here are some notable digital platforms for skill development:

i. **Coursera:** Coursera offers a vast array of online courses, specializations, and degree programs in collaboration with leading universities and institutions worldwide. Learners can access courses on topics ranging from computer science and data science to business, humanities, and beyond. Coursera's platform features interactive video lectures, quizzes, assignments, and peer-reviewed assessments.

ii. **edX:** Founded by Harvard University and MIT, edX is an open online course provider offering courses from top universities, colleges, and organizations worldwide. Learners can choose from a diverse selection of courses in fields such as computer science, engineering, healthcare, business, and more. edX offers both self-paced courses and instructor-led programs.

iii. **Udemy:** Udemy is a global online learning and teaching marketplace, featuring a vast library of courses on diverse topics taught by expert instructors. With Udemy, learners can access courses on everything from programming and web development to personal development, marketing, and creative skills. Courses are self-paced, allowing learners to study at their own convenience.

iv. **LinkedIn Learning:** LinkedIn Learning, formerly known as Lynda.com, is a professional development platform offering thousands of video courses taught by industry experts. Learners can access courses on business, technology, creative skills, and more, with a focus on practical, job-relevant skills. LinkedIn Learning also offers personalized recommendations based on learners' interests and career goals.

v. **Skillshare:** Skillshare is an online learning community offering thousands of creative and entrepreneurial skills classes. From graphic design and photography to writing, marketing, and entrepreneurship, Skillshare provides a platform for both learners and teachers to share their expertise and learn from each other through project-based learning.

vi. **Khan Academy:** Khan Academy offers free online courses, lessons, and practice exercises in subjects such as math, science, economics, history, and more. Targeted primarily at K-12 students, Khan Academy also provides college-level learning and test preparation resources. The platform features instructional videos, interactive exercises, and personalized learning pathways.

vii. **Codecademy:** Codecademy is an interactive platform offering coding courses and tutorials for beginners and experienced programmers.

Learners can access courses on programming languages such as Python, JavaScript, HTML/CSS, and more. Codecademy's hands-on approach allows learners to practice coding in real time and build projects to reinforce their skills.

These digital platforms for skill development provide learners with access to high-quality educational content, interactive learning experiences, and opportunities for continuous growth and advancement. Whether pursuing career-specific skills, exploring new interests, or seeking personal enrichment, these platforms offer many resources to support learners on their educational journey.

6.6.4 Skill Development Programs in India: An Overview

India has witnessed a significant expansion of skill development programs aimed at equipping its vast workforce with the skills needed to thrive in a dynamic global economy. Initiatives such as the Pradhan Mantri Kaushal Vikas Yojana (PMKVY), National Skill Development Corporation (NSDC), Skill India Mission, and state-level skill development programs have been instrumental in providing training and certification across various sectors, including manufacturing, healthcare, IT, hospitality, and agriculture.

Governments worldwide often implement various initiatives and programs to promote skill development among their citizens. These initiatives aim to address unemployment, improve workforce productivity, enhance employability, and foster economic growth. Here are some examples of government initiatives for skill development:

i. **Pradhan Mantri Kaushal Vikas Yojana (PMKVY):** Launched by the Government of India, PMKVY is one of the flagship programs aimed at providing skill training to youth across the country. It offers short-term skill training courses in various sectors, including manufacturing, retail, healthcare, construction, and more. PMKVY aims to certify over 10 million youth with industry-relevant skills by 2022 and promote entrepreneurship among trained individuals.

ii. **National Skills Development Corporation (NSDC):** NSDC is a public-private partnership organization established by the Government of India to promote skill development initiatives in the country. It partners with various training providers, industry associations, and employers to facilitate skill training programs, certification, and job placements. NSDC focuses on priority sectors such as automotive, construction, IT/IT, healthcare, and textiles.

iii. **SkillsFuture, Singapore:** SkillsFuture is a national movement in Singapore aimed at promoting lifelong learning and skill development among citizens. The initiative offers various programs and resources to support individuals in upgrading their skills, pursuing further education, and advancing their careers. SkillsFuture provides subsidies, grants, and training allowances to encourage participation in skill development activities.

iv. **European Social Fund (ESF), European Union:** The ESF is the European Union's main financial instrument for promoting employment and social inclusion through skill development and training initiatives. It supports a wide range of projects aimed at upskilling and reskilling individuals, particularly those facing barriers to employment, such as youth, migrants, and people with disabilities. ESF-funded programs include vocational training, apprenticeships, language courses, and job placement services.

v. **Workforce Innovation and Opportunity Act (WIOA), United States:** WIOA is a federal legislation in the United States aimed at promoting workforce development, job training, and education. It provides funding for various programs administered by state and local workforce development boards, including adult education, vocational rehabilitation, job training for dislocated workers, and apprenticeship programs. WIOA emphasizes collaboration among government agencies, employers, and education providers to meet the skill needs of employers and workers.

These are just a few examples of government initiatives for skill development, and many other countries have similar programs tailored to their specific needs and priorities. Overall, government interventions play a crucial role in fostering a skilled and adaptable workforce, driving economic growth, and promoting social inclusion.

6.6.5. Integration of Digital Literacy in Skill Development Programs

Recognizing the transformative potential of digital technologies, skill development programs in India have increasingly embraced digital literacy initiatives. These initiatives aim to enhance participants' proficiency in utilizing digital tools and platforms for learning, communication, and employment. Digital literacy components are integrated into training modules to familiarize participants with essential digital skills such as basic computer operations, internet navigation, email communication, and digital content creation.

6.7 Impact and Challenges

Integrating digital literacy into skill development programs has yielded tangible benefits, including increased employability, enhanced productivity, and greater economic opportunities. However, several challenges persist, including the digital divide, infrastructure constraints, and the need for continuous upskilling to keep pace with technological advancements. Addressing these challenges requires a multi-stakeholder approach involving government agencies, educational institutions, industry partners, and civil society organizations.

Lack of access to digital infrastructure, especially in rural regions, is one of the main problems. In contrast to 70% of urban homes, just 34% of rural families have access to the Internet in 2023, according to research. Due to the fact that many residents of rural regions lack access to digital services and cannot afford digital literacy training, the digital gap is a significant obstacle to digital literacy. Another issue is that some groups of people, especially women and elderly folks, are not aware of or interested in digital technology. In India, just 29% of women and 59% of males are technologically literate, according to data from 2023. In a similar vein, whilst 60% of persons between the ages of 18 and 29 are digitally literate, just 30% of adults over the age of 50 are. This emphasizes the necessity of focusing on raising these populations' level of digital literacy. In addition, the educational system has to place more of a focus on digital literacy. Many Indian schools and universities do not provide sufficient instruction in digital literacy, despite the rising significance of digital skills in the labor market.

6.8 Conclusion

In our ever-changing environment, the need of digital literacy and skill development cannot be emphasized. Understanding, using, and managing digital technologies is now essential as technology continues to infiltrate every part of our life. People who are digitally literate are more equipped to assess information critically, make wise decisions, and engage actively in the digital society. It goes beyond just knowing how to operate a computer or smartphone to comprehend the wider effects of technology on society and our daily life. The ability to use digital tools is becoming increasingly necessary for career success. Gaining expertise in fields like digital communication, data analysis, and coding guarantees adaptability in a labor market that is always evolving and offers doors to a variety of employment prospects. There are obstacles in the way of obtaining widespread digital literacy and skill development, though. Inequalities in access to high-quality education and technology continue to exist, aggravating already existing disparities. To guarantee fair access to digital resources and opportunities, governments, educational institutions, and other stakeholders must work together to address these discrepancies.

References

Adams, A. (2018). Social Media and Society: An Introduction. Cambridge University Press.

Ally, M. (2004). Foundations of educational theory for online learning. In Theory and practice of online learning (pp. 15-44). Athabasca University Press.

Brown, B. (2017). Cybersecurity: Protecting Critical Infrastructures from Cyber Attack and Cyber Warfare. CRC Press.

Chen, C. (2020). Mobile Commerce: Concepts, Methodologies, Tools, and Applications. IGI Global.

Clark, D. (2019). Cybersecurity Program Development for Business: The Essential Planning Guide. Wiley.

Czerniewicz, L., Deacon, A., Glover, M., & Walji, S. (2019). Online learning in higher education: A review of research on interactions among teachers and students. In The International Encyclopedia of Higher Education (pp. 1-8). Springer, Cham.

European Commission. (2019). Digital Education Action Plan: Digital literacy. Retrieved from https://ec.europa.eu/education/education-in-the-eu/digital-education-action-plan_en

Garcia, E. (2018). The Evolution of the Internet. ABC-CLIO.

Hernandez, H. (2020). Digital Government: Leveraging Innovation to Improve Public Sector Performance and Outcomes for the 21st Century. Rowman & Littlefield.

International Telecommunication Union (ITU). (2021). The State of Broadband: Broadband as a Foundation for Sustainable Development. Retrieved from https://www.itu.int/en/ITU-D/Statistics/Pages/publications/broadband.aspx

Johnson, K. (2021). E-Government 2.0: A Path Towards Sustainable, Citizen-Oriented Public Services. Springer.

Jones, J. (2019). The Dark Side of Social Media: Psychological, Managerial, and Societal Perspectives. Academic Press.

Kozinets, R. V. (2010). Netnography: Doing ethnographic research online. Sage.

KPMG. (2020). The Indian Technology Sector: Towards Global Leadership. Retrieved from https://assets.kpmg/content/dam/kpmg/in/pdf/2020/06/the-indian-technology-sector-towards-global-leadership.pdf

Kukulska-Hulme, A., & Traxler, J. (2007). Designing for mobile and wireless learning. Routledge.

Lee, L. (2019). Handbook of Research on Managing and Influencing Consumer Behavior. IGI Global.

Livingstone, S., & Helsper, E. J. (2007). Gradations in digital inclusion: Children, young people and the digital divide. New Media & Society, 9(4): 671-696.

Mayer, R. E. (2014). The Cambridge handbook of multimedia learning. Cambridge University Press.

Miller, M. (2020). Apps: From Concept to Consumer. ABC-CLIO.

Nguyen, N. (2018). Handbook of Research on Intelligent Techniques and Modeling Applications in Marketing Analytics. IGI Global.

Patel, P. (2021). Web Browsers: A Look at Chrome, Safari, Firefox, Edge, and Internet Explorer. Enslow Publishing.

Rheingold, H. (2012). Net smart: How to thrive online. MIT Press.

Roberts, R. (2019). The App Generation: How Today's Youth Navigate Identity, Intimacy, and Imagination in a Digital World. Yale University Press.

Sanchez, S. (2021). E-Government Strategy, ICTs, and Innovation for Citizen Engagement. IGI Global.

Smith, S. (2020). The Power of the Internet: Putting Information at Our Fingertips. Rosen Publishing.

Taylor, T. (2020). Digital Government: Overcoming the Challenges of Digital Transformation. Springer.

UNESCO. (2017). Media and Information Literacy Curriculum for Teachers. Retrieved from https://en.unesco.org/news/media-and-information-literacy-curriculum-teachers-now-available-online

UNESCO. (2020). Global Education Monitoring Report 2020: Inclusion and Education: All Means All. Retrieved from https://en.unesco.org/news/new-report-reveals-digital-divide-likely-widen-post-covid-19-pandemic

Weller, M. (2014). The battle for open: How openness won and why it doesn't feel like victory. Ubiquity Press.

White, W. (2018). Mobile Applications: Architecture, Design, and Development. CRC Press.

Wong, L. H., & Looi, C. K. (2011). What seams do we remove in mobile-assisted seamless learning? A critical review of the literature. Computers & Education, 57(4): 2364-2381.

7

Mobile Revolution Digital Literacy and Skill Development

Harshit Paliwal[1] and Pankaj Kumar[2]

[1]*Institute of Agricultural of Science, Banaras Hindu University, Varanasi Uttar Pradesh, India*

[2]*Department of Agricultural Extension, Chandra Shekhar Azad University of Agriculture and Technology, Kanpur, Uttar Pradesh, India*

Abstract

This chapter explores the transformative potential of mobile technology in the context of digital literacy and skill development. The pervasiveness of mobile devices has democratized access to digital resources, presenting immense opportunities for individuals and societies. It delves into foundational frameworks for digital literacy, outlining key competencies for both learners and educators. It then examines various strategies to foster digital literacy across educational settings, ranging from K-12 to lifelong learning initiatives. However, the chapter acknowledges the challenges that accompany the mobile revolution. The digital divide, rapid technological advancements, and concerns around misinformation and online safety necessitate comprehensive digital literacy education. The chapter proposes future research directions that can guide further exploration in this domain. These include longitudinal studies, cross-cultural comparisons, and investigations into the impact of emerging technologies on digital literacy.

Keywords: *Mobile Technology, Digital Literacy, Skill Development, Education, Empowerment.*

7.1 Introduction

The evolution of the digital landscape has been nothing short of revolutionary, shaping the way we live, work, and interact with one another. Since the inception of the ARPANET military experiment in 1968, which laid the groundwork for the internet as we know it today, the world's digitization has been accelerating at an unprecedented pace. What started as a military endeavor has evolved into a global network connecting billions of people across geographical boundaries, giving rise to virtual communities, and fundamentally transforming the way information is accessed and disseminated. As we navigate through this digital age, characterized by rapid technological advancements and an ever-expanding digital ecosystem, the role of individuals in this landscape has evolved significantly. Initially passive consumers of digital content, users have now become active participants, shaping and contributing to digital culture in various ways. This transformation, often described as the shift towards a "prosumer" model in participatory culture, underscores the dynamic nature of digital engagement (Toffler, 1980; Jenkins, 2005). However, alongside the opportunities presented by the digital revolution, there are also growing concerns regarding its impact on society. From the rise of phenomena like "democracy" to challenges posed by misinformation and cyber security threats, the digital landscape presents a complex array of issues that demand careful consideration (Fernandez-Villanueva and Bayarri Toscano, 2020; Salaverria *et al.*, 2020; Coventry and Branley, 2018).Moreover, the rapid pace of technological advancement often outstrips society's capacity to respond effectively, leading to a volatile, uncertain, complex, and ambiguous world, often referred to as a "V.U.C.A world" (Kingsinger and Walch, 2012). In this context, the development of digital skills becomes imperative, not only for personal empowerment but also for navigating the complexities of the digital age. Recognizing the critical importance of digital literacy and skill development, organizations and governments worldwide have renewed educational paradigms to equip individuals with the necessary competencies for the 21st century. The concept of "21st-century skills" has emerged as a holistic framework encompassing a range of competencies essential for effective participation today (Ananiadou and Claro, 2009; Voogt and Roblin, 2012). These skills go beyond mere technical proficiency, emphasizing critical thinking, creativity, and adaptability in the face of rapid change. With the proliferation of mobile technology, access to digital resources and opportunities has become more ubiquitous than ever before. However, this chapter explores how mobile technology intersects with digital literacy and skill development, examining its implications for education, workforce readiness, and societal empowerment.

Fig. 7.1 : This image represents the use of technology for education, a crucial aspect of digital literacy.

7.2 Digital Literacy Frameworks for Learners

Digital Competence Framework for Citizens (DigComp): Developed by the European Commission, DigComp serves as a foundational framework for many DL initiatives worldwide. It encompasses five competence areas, including information literacy, communication, digital content creation, safety, and problem-solving.

i. **Digital Literacy Global Framework (DLGF):** Developed by UNESCO, DLGF expands on DigComp by incorporating additional competence areas such as hardware and software operations, career-related competencies, and computational thinking.

ii. **ISTE Standards for Students:** Originating from the United States, these standards focus on empowering students with skills such as digital citizenship, innovative design, computational thinking, and global collaboration.

iii. **DL Frameworks in UK Higher Education:** Various frameworks in the UK emphasize academic practice, information literacy, media literacies, and ICT skills, reflecting a diverse approach to DL development.

iv. **Selber's Multiliteracies for a Digital Age:** Selber's framework emphasizes three student positions toward technology: users, producers, and questioners, with corresponding literacy filters including functional, critical, and rhetorical literacy.

7.3 Digital Literacy Frameworks for Educators

i. **ISTE Standards for Educators:** From the US, these standards delineate competencies for educators in areas such as leadership, collaboration, design, and analysis.

ii. **UNESCO's ICT Competency Framework for Teachers:** This framework outlines six aspects of teacher professional practice essential for integrating ICT into education effectively.

iii. **DigCompEdu:** Developed by the European Commission, this framework details 22 competencies across six areas, guiding educators in professional engagement, digital resources, teaching and learning, assessment, and empowering learners.

iv. **ENLACES:** Originating from Chile, ENLACES provides standards for teaching professionals focusing on educational, technical, management, social and ethical, and professional development competencies.

v. **Pozos Perez and Tejada Fernandez's Competencies:** These competencies highlight the diverse skills needed by university teachers, including planning, collaborative learning, pedagogical innovation, evaluation, and responsible ICT use.

7.4 Strategies to Foster Digital Literacy in Education

i. **K-12 and Secondary Education:** Strategies include promoting critical DL through explicit scaffolding, incorporating digital comics into curriculum, and training teachers to design and evaluate digital learning resources critically.

ii. **Higher Education:** Strategies involve integrating DL into curriculum design, providing opportunities for authentic learning experiences, supporting self-regulated learning, and leveraging librarian support for DL development.

iii. **Vocational Education:** Initiatives focus on embedding digital competencies into training delivery, enhancing teacher digital competence, and developing task-based online courses to support VET teachers.

iv. **Continuing Education:** Strategies emphasize the continuous development of digital competencies for educators, balancing technical and pedagogical knowledge, and promoting metacognition about competencies.

v. **Lifelong Learning:** Initiatives aim to address the diverse needs of lifelong learners, including adults, by providing tailored training programs, raising awareness of DL importance, and integrating DL into various aspects of daily life and work.

7.5 Challenges in Digital Literacy

i. **Digital Divide:** Despite the widespread access to mobile technology, disparities in digital access and literacy persist across socio-economic, geographic, and demographic lines. Bridging the digital divide remains a formidable challenge, requiring concerted efforts to ensure equitable access to digital resources and opportunities for all.

ii. **Technological Obsolescence:** The rapid pace of technological advancement poses challenges in keeping digital literacy skills relevant and up-to-date. Individual's risk falling behind as new technologies emerges, highlighting the need for continuous learning and adaptation in the digital age.

iii. **Misinformation and Digital Literacy:** The proliferation of misinformation and fake news on digital platforms underscores the importance of critical digital literacy skills. Addressing this challenge requires equipping individuals with the ability to critically evaluate and discern the credibility of online information.

iv. **Privacy and Security Concerns:** With increasing digitization come heightened concerns about privacy and cyber security. Individuals need to develop skills to protect their digital identities and navigate online spaces safely, highlighting the importance of incorporating digital security education into digital literacy initiatives.

v. **Digital Fatigue and Overload:** The constant bombardment of digital information and stimuli can lead to digital fatigue and overload, impacting individuals' well-being and productivity. Strategies to manage digital consumption and promote digital well-being are essential aspects of comprehensive digital literacy education.

7.6 Future Research Directions

i. **Longitudinal Studies:** Longitudinal studies tracking individuals' digital literacy development over time can provide valuable insights into the long-term impact of digital literacy interventions and identify factors that contribute to sustained digital competency.

ii. **Cross-Cultural Studies:** Comparative studies across different cultural contexts can shed light on the cultural nuances of digital literacy and inform culturally responsive digital literacy education initiatives tailored to specific communities' needs and contexts.

iii. **Intersectionality and Digital Literacy:** Exploring the intersectionality of factors such as gender, race, ethnicity, and socio-economic status in digital literacy can uncover disparities and inequities in access and opportunities, informing more inclusive and equitable digital literacy interventions.

iv. **Digital Literacy in Emerging Technologies:** With the advent of emerging technologies such as artificial intelligence, augmented reality, and block chain, investigating their implications for digital literacy and developing corresponding literacy frameworks is essential to prepare individuals for the future digital landscape.

v. **Innovative Pedagogies for Digital Literacy:** Research into innovative pedagogical approaches, such as gamification, project-based learning, and collaborative inquiry, can enhance digital literacy education effectiveness by engaging learners actively and fostering practical digital skills.

vi. **Digital Literacy Assessment:** Developing robust assessment tools and methodologies for measuring digital literacy competencies is crucial for evaluating the effectiveness of digital literacy interventions and informing evidence-based practice in digital literacy education.

7.7 Conclusion

In conclusion, the intersection of mobile technology with digital literacy and skill development presents a wealth of opportunities and challenges for individuals, educators, and societies at large. The evolution of the digital landscape has transformed the way we engage with information, communicate with others, and navigate the complexities of the modern world. From the foundational frameworks for learners and educators to the strategies aimed at fostering digital literacy across various educational settings, there is a concerted effort to equip individuals with the necessary competencies for the 21st century. However, amidst the opportunities presented by the mobile revolution, there are also significant challenges that need to be addressed. The persistent digital divide, concerns about technological obsolescence, misinformation, privacy and security issues, and digital fatigue underscore the importance of comprehensive digital literacy education. Looking ahead, future research directions offer avenues for deeper exploration and innovation in digital literacy. Longitudinal studies, cross-cultural comparisons, and investigations into emerging technologies will provide valuable insights into the evolving nature of digital literacy and inform evidence-based practices. Moreover, the development of innovative pedagogies and robust assessment tools will enhance the effectiveness of digital literacy education initiatives. In essence, as we continue to navigate the complexities of the digital age, fostering digital literacy and skill development remains imperative for personal empowerment, workforce readiness, and societal advancement. By addressing challenges, embracing opportunities, and advancing research in this field, we can cultivate a digitally empowered society capable of thriving in an ever-changing technological landscape.

References

Ananiadou, K. & Claro, M. (2009). 21st Century Skills and Competences for New Millennium Learners in OECD Countries. OECD Education Working Papers. https://doi.org/10.1787/218525261154

Bravo, M. C. M., Chalezquer, C. S., & Serrano-Puche, J. (2021). Meta-framework of digital literacy: A comparative analysis of 21st-century skills frameworks. Revista Latina de Comunicacion Social, (79): 76-109.

Coventry, L., & Branley, D. (2018). Cybersecurity in healthcare: A narrative review of trends, threats, and ways forward. Maturitas, 113: 48-52. https://doi.Org/10.1016/j.maturitas.2018.04.008.

Fernandez-Villanueva, C. & Bayarri Toscano, G. (2021). Legitimation of hate and political violence through memetic images: the Bolsonaro campaign. Communication & Society, 34(2): 449-468. Doi: 10.15581/003.34.2.449-468.

Jenkins, H. (2006). Convergence Culture: La cultura de la convergencia de los medios de comunicacion. Ediciones Paidos Iberica.

Kingsinger, P. & Walch, K. (2012). Living and leading in a VUCA world. Thunderbird University.

Slaveries, R., Buslon, N., Lopez-Pan, F., Leon, B., Lopez-Goni, I & Erviti, M.C. (2020). "Desinformacion en tiempos de pandemia: tipolog^a de los bulos sobre la Covid-19". El profesional de la informacion, v. 29, n. 3, e290315. https://doi.org/10.3145/epi.2020.may.15.

Toffler, A. (1980). The Third Wave. Vol. 484, Bantam books, New York, NY.

Villaroel, G. (2001). El metodo comparativo: entre complejidad y generalization. Revista Venezolana de Ciencia Polftica. v.20 (jul-dic2001) pp. 97-120.

Voogt, J. & Roblin, N. (2012). A comparative analysis of international frameworks for 21st century competences: Implications for national curriculum policies. Journal of Curriculum Studies, 44(3): 299-321. https://doi.org/10.1080/00220272.2012.668938

8

Financial Inclusion Fintech and Access to Banking Services

***Amrit Banerjee*[1] *and Deeptimayee Mohapatra*[2]**

[1]*College of Agriculture, Odisha University of Agriculture and Technology Bhubaneswar, India*

[2]*College of Agriculture, MITS Institute of Professional Studies, Rayagada Odisha, India*

Abstract

Combining the terms "financial" and "technology," we get "FinTech." It can be broadly defined as financial innovation enabled by technology that may lead to new applications, processes, business models, or products that significantly impact financial markets, institutions, and the provision of financial services. Therefore, fintech has the power to drastically alter the Indian financial services and financial inclusion landscape. FinTech companies can contribute to greater competition and the acceleration of Financial Inclusion in India by lowering costs and enhancing access to financial services for the underprivileged, low-income individuals, rural residents, and other underserved segments of the Indian economy through their innovations, new business models, and applications. Nearly 90% of micro units in India are thought to still be unaffiliated with the official credit system. This group and small businesses stand to gain greatly from a partnership between banks and FinTech companies, since their payment histories can be used to determine their creditworthiness. It is noteworthy that a number of FinTech companies have experienced rapid expansion in recent years; whereas at the start of this decade they were essentially unknown, there are currently as many as 1218 entities operating in India. This chapter mainly discusses the definition of financial inclusion, the role of financial inclusion in reaching a greater number of people, the role of fintech in financial inclusion, how the financial inclusion concept helps people get banking services and the benefits and challenges of financial inclusion.

Keywords: *Financial Inclusion, Fintech, Banking service.*

8.1 Introduction

Giving people and companies access to basic financial services is known as financial inclusion, and it has long been a problem on a global scale. Millions of people worldwide still lack access to basic banking services like savings accounts, loans, and insurance and are either underbanked or unbanked altogether. Fintech, or financial technology, has emerged as a key player in resolving this problem by providing creative ways to improve banking service accessibility and advance financial inclusion.

One of the main forces behind promoting equitable economic growth and lowering poverty is financial inclusion. Ensuring complete financial inclusion has been a top priority in the context of India, a nation with a diverse population that lives in large parts in underserved and rural areas. India has come a long way in this regard over the years, using policy changes and technological advancements to bring a greater proportion of the populace into the formal financial system.

According to a report by IFC and Intellecap, the formal banking sector provides funding for approximately 16% of MSMEs in India. In India, the vast majority of micro units remain unaffiliated with the official credit system. The MSME financing gap is estimated to be US $240 billion, while the consumer lending gap is estimated to be US $300 billion, according to the India FinTech Report 2019. Based on their other payment histories and other data, financial institutions and FinTech companies can work together to evaluate the credit worthiness of MSME and consumer loan segments, which stand to gain greatly from this collaboration. The numbers show that there is a significant gap between supply and demand.

FinTech lenders utilize two different credit scoring models than traditional lenders do. First, technology allows financial intermediaries to gather and utilize more data. FinTech credit platforms might use alternative data sources include user digital footprints (Berg *et al.*, 2018) and insights from social media activity (U.S. Department of the Treasury, 2016; Jagtiani and Lemieux, 2018a). The second distinction is the use of machine learning techniques, which can extract non-linear information from variables.

In light of current technological advancements, serving India's underprivileged populations also makes sense commercially. The Consultative Group to Assist the Poor (CGAP) claims that FinTechs are creating new value propositions at every stage of the financial services value chain, such as flexible products and more effective ways to handle the financial difficulties that low-income clients face. Because of this, India's adoption of FinTech has increased dramatically over the past two years, driven by regulatory and governmental initiatives to promote financial inclusion and a digital economy.

8.2 Financial Inclusion: A Revolutionary Fintech Solution in the Present Situation

Financial inclusion is a transformative force in the modern global finance landscape, arising from the intersection of technology and inclusivity. This paradigm shift has become a potent catalyst for addressing economic disparities and empowering people previously shut out of mainstream financial services, largely thanks to fintech innovations. In the current context, financial inclusion is a great fintech that is transforming how consumers interact with and obtain banking services.

i. **Mobile Banking and Accessibility:** Mobile banking's rapid growth is one of fintech's major contributions to financial inclusion. Due to the increasing availability of smartphones and internet connectivity, mobile banking has become a game-changer, particularly in areas with restricted access to traditional banking infrastructure. Nowadays, people have access to a wide range of financial functions, including transferring money and checking account balances. Due to its accessibility, banking services are now available to those who were previously underbanked or unbanked.

ii. **Digital Wallets:** Another fintech innovation that has been instrumental in empowering the unbanked is digital wallets. Thanks to these electronic systems, users no longer require traditional bank accounts to conduct transactions, store money securely, and engage in the formal economy. These days, digital wallets are a lifesaver for people in difficult financial situations because they offer a secure and practical way to handle money, especially in unpredictable economic times.

iii. **Blockchain Technology:** The underlying technology of cryptocurrencies like Bitcoin, blockchain technology, has expanded the scope of financial inclusion. Transactions are safe and resistant to tampering because of its transparent and decentralized structure. Blockchain is becoming a fintech solution that fosters trust in the current climate, where privacy and data security are major concerns. This is especially true when it comes to facilitating safe and transparent financial transactions.

iv. **Fintech for Micro-finance and small business:** Fintech solutions have made microfinance possible and are now a potent tool for financial inclusion. Peer-to-peer lending is facilitated by fintech platforms that link borrowers and lenders directly, making it easier for individuals and small enterprises to obtain credit. This is especially important given the current circumstances, at which small businesses' growth and resilience are key to the economy's recovery.

v. **RegTech:** Another fintech innovation simplifying compliance procedures for financial institutions is regulatory technology, or RegTech. RegTech solutions enable institutions to quickly adapt to constantly changing regulatory requirements, preventing compliance from impeding efforts to promote financial inclusion.

8.3 Financial Inclusion: Access to Banking Services

People benefit greatly from financial inclusion because it gives them access to necessary banking services. Ensuring that all people and businesses have access to and use a range of financial services is central to the idea of financial inclusion, which also aims to empower communities and encourage economic stability. Financial inclusion makes it easier for people to obtain banking services in the following ways:

i. **Access to banking services:** The goal of financial inclusion is to integrate people into the established financial system so they can open savings accounts, apply for credit, and receive insurance benefits. This access allows them to manage their money in a controlled and safe environment.

ii. **Facilitates savings and asset buildings:** People who are financially included are able to save money in official banking institutions. This promotes long-term financial stability by encouraging a culture of saving and asset building in addition to helping to secure their funds.

iii. **Access to credit and capital:** For those without collateral or a credit history, financial inclusion makes credit more accessible. This is crucial because it gives entrepreneurs and small companies the funding they need to launch or grow their enterprises.

iv. **Promotes entrepreneurship:** People who have access to banking services can more readily engage in entrepreneurial activities. They can manage transactions, take advantage of loans, and engage in the formal economy, all of which promote entrepreneurship and regional economic expansion.

v. **Enhances financial literacy:** Initiatives for financial inclusion frequently include educational components aimed at raising financial literacy. Through these programs, people can gain an understanding of the advantages and responsibilities of banking services, enabling them to make wise financial decisions.

vi. **Reduces reliance on informal channels:** Due to limited access to formal banking services, many people, particularly in underserved areas, rely on informal financial channels. Financial inclusion offers safer and more regulated alternatives that shield people from the risks associated with unregulated practices, which help decrease this reliance.

vii. **Promotes digital financial services:** Digital financial services are becoming increasingly popular because technology integrates with financial inclusion. Thanks to mobile banking, digital wallets, and other fintech solutions, people can manage their finances easily and conveniently without going to physical bank branches.

viii. **Ensures financial security:** The formal financial system affords a certain amount of financial security. Banking services frequently provide insurance options, safeguard funds, and guard against fraud, improving an individual's and their family's overall financial security.

ix. **Supports inclusive economic growth:** One of the main forces behind inclusive economic growth is financial inclusion. The availability of banking services to a greater proportion of the populace fosters a stronger and more resilient economy, which in turn benefits society as a whole.

8.4 Role of Fintech in Accelerating Financial Services in India

FinTech is helping to remove long-standing obstacles to financial inclusion and the deepening and development of the financial sector in India. The potential for digital financial services to help remove geographical barriers to financial inclusion is growing. FinTech solutions have been identified as potentially transformative in several areas, including financial inclusion. This is according to an IMF paper titled "Fintech: The Experience so Far" (PolicyPaperNo.19/024 published on June 27, 2019). This is because they help address several financial frictions. These comprise: (a) financial service delivery cost barriers, which are particularly severe in rural areas far from cities and among marginalized populations like women, the impoverished in cities, and migrants; (b) information asymmetries between service providers and customers, particularly among the unbanked who lack the knowledge necessary to evaluate risk; (c) inability to provide a verifiable ID and trouble fulfilling customer due diligence obligations; and (d) absence of appropriate financial products for lower income groups.

However, India's FinTech revolution has had the greatest impact on the MSME lending landscape. New and creative alternative lending platforms have made it possible for smaller businesses without credit histories or financial records to finally obtain much-needed credit. Additionally, the Goods and Services Tax Network (GSTN), which presently has over 9.2 million MSMEs registered, submits monthly returns on a regular basis. The matching concept verifies the GSTN data, which offers a more comprehensive and in-depth picture of the business's nature and supplements traditional financial data. Stronger underwriting procedures made possible by more precise transactional information verification and validation will increase the effectiveness and coverage of more recent FinTech lending models, like flow-based lending, vertical-based learning and ecosystem based lending. (Mittal, 2020)

FinTech have the potential to increase the number of financial service providers and create more inclusive and efficient business models. It's also commonly acknowledged that FinTech has immense potential to boost economic development and build a more equitable financial system. According to a McKinsey Global Institute report (Manyika, *et al.* 2016), if emerging nations completely adopt digital finance, their GDP might increase by $3.7 trillion by 2025. Additionally, studies have demonstrated that nations with greater financial inclusion had slower rates of economic growth and less income disparity (Ravi, Shamika 2019).

8.5 Benefits of Financial Inclusion

Financial inclusion, the process of guaranteeing that all people and businesses have access to necessary financial services, significantly influences socioeconomic development. The advantages of financial inclusion are becoming increasingly clear as countries work to develop inclusive economies. This essay examines the many benefits of promoting financial inclusion, from increased economic growth to personal empowerment.

i. **Poverty Alleviation:** Financial inclusion offers a way out of poverty for millions of people. Individuals who have access to formal financial services, like credit, insurance, and savings accounts, are better equipped to handle financial shocks, make investments in sources of income, and create a safety net for their finances. Families are able to break free from the cycle of poverty and work toward a safer and more prosperous future as a result.

ii. **Entrepreneurial Opportunities:** Financial inclusion facilitates entrepreneurship by giving people the resources to launch or grow small enterprises. Credit availability allows would-be business owners to finance investments, stock purchases, and business expansion. This promotes innovation and the creation of jobs, in addition to boosting local economies.

iii. **Women's Empowerment:** Financial inclusion transforms gender equality. Access to banking services is especially beneficial for women, who frequently have more financial constraints. Financial inclusion helps women become more economically empowered by allowing them to save, invest, and obtain credit. This improves their standard of living and increases their ability to make decisions for their households.

iv. **Stimulating Economic Growth:** Economic growth is facilitated by a populace that is inclusive of all income levels. Financial inclusion raises economic productivity through promoting savings, entrepreneurship, and easier access to credit. Small and medium-sized businesses (SMEs) thrive, generating jobs and promoting a more resilient and dynamic economy.

v. **Reducing Income Inequality:** Reducing income inequality largely depends on financial inclusion. It contributes to closing the wealth gap by giving low-income and marginalized populations access to formal financial services. The advantages of economic growth are distributed more fairly when everyone has the chance to engage in the formal financial system.

vi. **Enhancing Education & Healthcare:** Families can invest in healthcare and education when they are financially included. People who have access to credit and savings can pay for healthcare needs as well as educational costs like books and school fees. By improving human capital, this promotes a population that is healthier and better educated.

vii. **Digital Transformation and Innovation:** Financial inclusion has been made possible in large part by the development of fintech, or financial technology. The delivery of financial products is made possible by digital financial services like digital wallets and mobile banking, which are more economical and efficient. This shift to digital technology not only makes life easier for users, but it also encourages financial inclusion on a larger scale.

8.6 Opportunities in Financial Inclusion

i. **Digital Onboarding and Financial Inclusion:** Reaching the goal of financial inclusion requires digital onboarding. Since almost 100 million KYC records have already been uploaded to this platform, the central KYC registry represents a significant step in this direction. Currently, "digital financial inclusion"—which broadly refers to using digital financial services to advance financial inclusion—is the focus of attention. These consist of digitally delivered payments, transfers, savings, credit, insurance, securities, etc.

ii. **Designing suitable financial products:** FinTech is essential to making financial platforms more accessible in the Indian context. To achieve the goal of financial inclusion, it is therefore essential to design appropriate financial products that address the unique needs of the financially excluded population.

iii. **Eliminating the gender gap:** Women's empowerment and financial inclusion can have profound socioeconomic effects. According to the World Bank's Global Findex 2017 Report, men are more likely to have an account than women. Furthermore, data gaps based on gender can keep women at a disadvantage. Gender data can therefore be gathered in accordance with the High-Level Committee on Deepening of Digital Payments' recommendation in order to quantify and ultimately close this gap. FinTech can play a significant role in empowering women by creating financial products and services tailored to their needs and making them more accessible to them via the internet, mobile devices, and personal computers, among other platforms.

iv. **Bringing merchants in the digital fold:** The great majority of Indian merchants are still not using digital platforms. Only a small percentage of the seven crore traders and merchants have started to accept digital payments, so there is a lot of room to grow merchant acceptance through QR codes, mobile payments, and smart point of sale (PoS) terminals. This would be feasible if merchants also received services other than payments, like lending, reconciliation, working capital, etc. Enhancing the infrastructure and connectivity for accepting payments in regions with minimal digital transactions would also significantly boost financial inclusion in India.

8.7 Risks

RegTech and SupTech: Regarding potential risks associated with FinTech adoption and how to mitigate them, emerging industries like RegTech and SupTech can help boost productivity by automating processes, introducing new features, and optimizing work flows. RegTech is an application or platform that reduces the cost of compliance and increases the efficiency of regulatory compliance through automated processes. RegTech is centered on technologies that make it easier and more efficient to deliver regulatory requirements. Technology that regulators and supervisors can use to support supervision is known as "SupTech." Data collection and reporting, data analysis and decision making, simplified licensing, market monitoring and surveillance, KYC, AML, and CFT, cyber security data, and evidence-based policy making are all goals of SupTech.

SupTech is being used by the Reserve Bank of India to collect and analyze data. A few of the examples are the Central Repository of Information on Large Credits (CRILC), the Export Data Processing and Monitoring System (EDPMS), and the Import Data Processing and Monitoring System (IDPMS). Additionally, one example of SupTech is the risk-based supervision of banks, which is heavily data-driven. However, big data analytics, AI, machine learning, cloud computing, data transfer protocols, biometrics, geographic information system (GIS) mapping, and other technologies are where RegTech and SupTech technologies will be found in the future.

8.8 Challenges in Financial Inclusion

8.8.1 Lack of Infrastructure

i. **Challenge:** Banks, ATMs, and internet access are examples of the digital and physical infrastructure that is lacking in many developing nations.

ii. **Strategy:** Make the necessary infrastructure investments, such as enlarging banking networks, enhancing connectivity, and advancing electronic payment methods.

8.8.2 Limited Financial Literacy

i. **Challenge:** Many people lack financial literacy and are ignorant of fundamental financial concepts, particularly in rural areas.

ii. **Strategy:** Put in place educational initiatives to improve financial literacy, addressing subjects like saving money, budgeting, and comprehending financial products.

8.8.3 Low Income and Unstable Payment

i. **Challenge:** People with irregular incomes or unstable employment may find it challenging to obtain financial services and to take advantage of them.

ii. **Strategy:** Create adaptable financial products, like pay-as-you-go and microfinance, that are suited to erratic income patterns.

8.8.4 Regulatory Barriers

i. **Challenge:** Tight regulations have the potential to impede the advancement of creative financial solutions and restrict the accessibility of financial services.

ii. **Strategy:** Support inclusion initiatives by collaborating with regulators to establish a regulatory framework that fosters financial innovation and is appropriate in size.

8.8.5 Social and Cultural Factors

i. **Challenge:** Certain demographic groups, especially women, may be discouraged from participating in formal financial systems by social and cultural norms.

ii. **Strategy:** Develop financial services and products that consider cultural sensitivity and launch awareness campaigns to dispel prejudice and advance inclusivity.

8.8.6 Limited Access to Identification

ii. **Challenge:** People may not be able to access financial services or open bank accounts if they do not have the necessary identity documents.

ii. **Strategy:** Encourage the adoption of digital identity solutions and work with authorities to enhance the issuance of identity documents.

8.8.7 Technological Barriers

i. **Challenge:** Insufficient availability of computers, smartphones, and the internet may impede the uptake of digital financial services.

ii. **Strategy:** Promote the creation of accessible and inexpensive technology, and extend the network's infrastructure to underserved regions.

8.9 National Strategy for Financial Inclusion (NSFI)

The Reserve Bank of India defines Financial Inclusion as the *"process of ensuring access to appropriate financial products and services needed by all sections of the society in general and vulnerable groups such as weaker sections and low income groups in particular, at an affordable cost in a fair and transparent manner by regulated, mainstream institutional players"*. (Jain M.K, 2019). Formal financing access can improve investments in human capital, decrease susceptibility to economic shocks, and promote the development of jobs. Macroeconomically speaking, increased financial inclusion can promote equitable and sustainable socioeconomic progress for all. The Reserve Bank of India has also created a National Strategy for Financial Inclusion for India 2019-24 in order to accomplish the aforementioned goals in a coordinated and timely way. In order to promote inclusive and resilient multi-stakeholder-driven growth, the Strategy aims to make formal financial services available, accessible, and affordable to all residents in a safe and transparent way. (NSFI: 2019-2024).

Table 8.1. A snapshot of the aspect and Descriptions of the National Strategy for Financial Inclusion (NSFI):

Aspect	Description
Goal	Make formal financial services accessible and affordable to all Indians.
Vision	"A financially aware and empowered India"
Focus	Reaching vulnerable groups and low-income segments
Benefits	Increased investment in human capital - Reduced vulnerability to economic shocks - Job creation - Equitable and sustainable socioeconomic progress
Key objectives	Universal access to financial services
Created by	Reserve Bank of India (RBI)
Timeframe	2019-2024

8.10 The Next Wave of Financial Inclusion

By enabling people to create digital and financial identities and proving their creditworthiness and eligibility for various financial services previously unavailable to them, data analytics, machine learning, and alternative data sources are paving the way for the next wave of inclusion. Biometrics and e-KYC are helping the unbanked population advance and support the next wave of inclusion (Dunaev, 2018). The widespread use of India's biometric linked identification, Aadhaar, has greatly sped up the KYC procedures, making creating bank accounts and electronic wallets easier. Rural residents may make deposits, withdrawals, and transfers using the Aadhaar-enabled Payment System (AePS). Business correspondents (BCs) can visit remote regions with a micro ATM that can verify customer's identity digitally and operate their accounts. (Carstens, Agustin 2019).

8.11 Conclusion

Fintech-driven financial inclusion can improve millions of lives by giving people access to basic banking services. Peer-to-peer lending, digital wallets, blockchain technology, and mobile banking enable people and businesses, especially in areas where traditional banking is scarce. The convergence of fintech and financial inclusion will be essential to building a more just and inclusive global economy as long as regulations change and technology advances. Governments, financial institutions, and fintech companies are among the stakeholders who must work together to overcome obstacles and create a future where everyone can fully participate in the financial system.

References

Anonymous. (2017). The Global Findex Database, retrieved from https://globalfindex. worldbank. org/

Anonymous. (2018). FinTech for Inclusion: Challenges and Opportunities, Roundtable discussion organized by Business Today in collaboration with PwC, BT Print Edition: December 16, 2018. Retrieved from https://www.businesstoday.in/magazine/the-hub/fintech-for-inclusion-challenges-and-opportunities/story/294454.html.

Anonymous. (2018). What's Driving India's Fintech Boom?, Feb.11,2018. Retrieved from https://knowledge.wharton.upenn.edu/article/whats- driving-indias-fintech-boom

Anonymous. (2019). India FinTech Report, retrieved from https://mediciinnercircle.com/wpcontent/uploads/2019/03/FintegrateReport_ExecutiveSummary_Final.

Anonymous. (2019). Report of the Steering Committee on Fintech Related Issues 2019, Department of Economic Affairs, Ministry of Finance, Government of India, accessed on January 09, 2020

Anonymous. (2019). Fintech: The Experience So Far IMF Policy Paper Policy Paper No. 19/024, retrived from https://www.imf.org/en/Publications/Policy-Papers/Issues/2019/06/27/Fintech-The-Experience-So-Far

Anonymous. (2019). Global Fintech Survey, retrieved from https://www.pwc.com/gx/en/industries/financial-services/fintech-survey/blurred-lines.html

Anonymous. (2019). EY Global FinTech Adoption Index 2019, retrived from https://www.ey.com/en_gl/ey-global-fintech-adoption-index, accessed on January 09,2020

Anonymous. (2020). FinTech in India- Ready for breakout, Report by Deloitte, retrieved from https://www2.deloitte.com/in/en/pages/financial-services/articles/fintech-india-ready-for-breakout.

Anonymous (2020).https://economictimes.indiatimes.com/small-biz/sme-sector/msmes-indias-growth-engine-face-a-rs-16-lakh-crore-credit-gapmsmeday/articleshow/69968156.cms, accessedon January 09, 2020

Anonymous(2020).https://www.cgap.org/research/publication/fintechs-and-financial-inclusion, accessed on January 09, 2020

Bansal S. (2019). Innovate & differentiate for true financial inclusion, retrived from https://www.livemint.com/companies/people/innovate-differentiate-for-true-financial-inclusion-sachin-bansal-11575484235987.html, accessed on January 09, 2020

Carstens A. (2019). Central Banking and Innovation: Partners in the Quest for Financial Inclusion, The seventeenth C.D. Deshmukh Memorial Lecture Delivered by Mr. Agustm Carstens, General Manager, Bank for International Settlements on April 25, 2019

Das S. (2019). Opportunities and Challenges of FinTech, Shri Shaktikanta Das, Governor, Reserve Bank of India, Keynote Address Delivered at the NITI Aayog's FinTech Conclave on March 25, 2019

Dunaev A. (2020). Governments key to solving financial inclusion, retrived from https://www.finextra.com/blogposting/16391/governments-key-to-solving-financial-inclusion.

Gambacorta L. *et al.* (2019). How do machine learning and non-traditional data affect credit scoring? New evidence from a Chinese fintech firm, BIS Working Papers No 834.

Jagtiani J. and Lemieux C. (2017). "Fintech lending: Financial inclusion, risk pricing, and alternative information".

Jagtiani, J. and Lemieux C. (2018). "Do fintech lenders penetrate areas that are underserved by traditional banks", Federal Reserve Bank of Philadelphia Working Papers, 18-13, March.

Jagtiani, J. and Lemieux C. (2018). "The roles of alternative data and machine learning in fintech lending: evidence from the Lending Club consumer platform", Federal Reserve Bank of Philadelphia Working Papers, 18-15, April

Jagtiani J., Lambie-Hanson L. and Lambie-Hanson T. (2019). "Fintech lending and mortgage credit access", Federal Reserve Bank of Philadelphia Working Papers, 19-47, November.

Jain M. K. (2019). Micro finance as the next wave of Financial Inclusion, Address by Shri M.K. Jain, Deputy Governor, Reserve Bank of India at the SIDBI National Microfinance Congress 2019 at Mumbai on November 26, 2019

Manyika J. *et al.* (2016). Digital Finance for All: Powering Inclusive Growth in Emerging Economies, McKinsey Global Institute, September 2016.

Misra A. and Tankha A. (2019). Inclusive Finance India Report, 2019

Mittal A. (2020). From small to big: The evolution of MSME lending in 2020 and the role of fintech, January 2020, retrived from https://economictimes.indiatimes.com/small-biz/sme-sector/from-small-to-big-the-evolution-of-msme-lending-in-2020-and-the-role-of-fintech/articleshow/73044876.cms?from=mdr,accessed on January09, 2020

Rao S. K. (2019). 'Digital literacy is integral for financial inclusion', published on December 30, 2019 at https://www.thehindubusinessline.com/opinion/digital-literacy-is-integral-for-financial-inclusion/article30432071.

9

Telehealth and E-Learning: Expanding Access to Healthcare and Education

***Rohit*[1], *Kumari Asha*[2], *Ashutosh Upadhyay*[3] and *Pranjal Singh*[3]**

[1]*College of Agriculture, Lords University, Alwar, Rajasthan, India*

[2]*G.L.M. Agriculture college, Kishangarh bas, Alwar (Affiliated to S.K.N. Agriculture University, Jobner, Jaipur), Rajasthan, India*

[3]*Chandra Sekhar Azad University of Agriculture & Technology, Kanpur Uttar Pradesh, India*

Abstract

Telehealth and e-learning have become known as innovative movements in healthcare and education, respectively, reshaping how services are given and knowledge is shared. This chapter investigates the interaction connecting telehealth and e-learning, emphasizing their complementary potential to enhance accessibility, effectiveness, and quality in both medical services and educational activities. Telehealth services provide a wide range of healthcare services, such as basic medical care, psychological counseling, and specialized consultations, allowing individuals to receive comfortable and timely healthcare irrespective of their location. At the same time, e-learning has transformed education by making flexible and active learning environments available to learners globally. Learners can use online platforms to access a wide range of educational materials, participate in realistic simulations, and participate in learning activities. However, issues such as digital literacy, concerns about privacy, and discrepancies in access to technology need to be addressed in order to promote fair access to and utilization of telehealth and e-learning tools. Telehealth and e-learning offer excellent opportunities to improve healthcare delivery and higher education, encouraging people, medical professionals, and learners to succeed in an interrelated world of technology while also improving the dissemination of information globally.

Keywords: *Telehealth, e-learning, healthcare, education*

9.1 Introduction

In the 21st century, mass media plays an important role in disseminating quick information among many people and making communication easy. Today, Information and Communication Technology (ICT) plays a significant role in daily life. ICT is the abbreviation for Information Technology (IT), which offers the ability to access information using telecommunication. It is primarily dependent on communication technologies. This incorporates computer system, sound video show and Internet. Information and communication technology is used in the majority of fields, for example, Agriculture, Education, Medicine, Defence, E-administration, E-Commerce, Banking, Transport, and so forth. ICT is assuming an essential job. Teleeducation, often known as e-education, refers to remote education that uses information and communication technology (ICT), as well as ongoing beneficiary education and the use of digital libraries, databases, or electronic data with knowledge bases. Distance learning (E-learning) is an element of tele-education has grown in popularity over the last decade; nevertheless, its implementation varies greatly among medical schools and seems to be more widespread in basic scientific courses than in clinical teaching.

1.2 Telehealth

In an era of growing demands on medical worker efficiency, technological advances in communication and computing enable health professionals to experiment with a variety of applications that may allow them to satisfy medical needs while also contributing to research initiatives. Between 1960 and 2000, telehealth was increasingly employed to provide healthcare to communities in underserved and difficult-to-access places worldwide, such as rural and village populations, incarcerated populations, etc.

Telehealth may be unfamiliar to certain healthcare workers; however, it is not new. Pedal radios have been used for remote medical care since the 1920s, when the "Royal Flying Doctor" assistance in Australia was founded. NASA supported the growth of telehealth by funding research into novel ways to provide healthcare to astronauts in space, as well as in-flight health care. Telehealth can be characterized as the application of digital or electronic information and telecommunication technology for supporting extended distances professional health care, patient and practitioner, medical education, public wellness, and health administration. Examples of technology include teleconferencing, accessibility to the internet, visual storage and forwarding, multimedia streaming media, and landline and wireless communications.

Telemedicine was defined in 1996 by the Institute of Medicine in the United States as the use of digital technology for communication and information to deliver and assist medical care where individuals are geographically separated by location.

Fig. 9.1: Shows On-Demand Telemedicine

9.2.1 Challenges in Telehealth

i. One concern is information security, particularly when it comes to protected patient information on telehealth systems. Several factors, including data encryption, user identification and authorization, secure access controls, and others, must be addressed to maintain the security, integrity, and availability of patient information in telehealth systems. Managing these information security challenges is critical to supporting both patient and healthcare provider trust in telehealth systems while also ensuring the confidentiality and safety of sensitive health information.

ii. The second question is how telemedicine affects the medical profession across national and international boundaries. Telehealth removes all physical boundaries between patients and healthcare practitioners, necessitating significant changes to clinical licensure regulations that have previously been based on geographic borders. Language diversity, limited physical inspections, and other factors all significantly impact telehealth. Limitations in conducting full physical examinations and communication between healthcare workers and patients of varied linguistic origins may impact diagnostic accuracy.

9.2.2 Classification of Telehealth

Based on the delivery mechanism, telehealth is classified as follows:

i. **Asynchronous telehealth:** Since telemedicine services can use primarily asynchronous or interactive methods, the application itself determines the system used. Asynchronous modes have been employed to facilitate medical evaluations without simultaneous patient-doctor interaction. Phone conversations and emails, for example, are often the first apps physicians use to enhance medical services and facilitate communication whenever their schedules render them challenging to contact. These are used to answer patients' inquiries, improve patient awareness, supplement program screening, and enhance compliance with treatment guidelines. Some medical professionals have used email to interact with patients, but the majority believe it could be excessively time-consuming. In contrast to practically each national survey, many medical professionals feel that a large proportion of their patients have less interest in connecting with them through the internet, significantly underestimating their patients' use of the internet. Responding to an email typically requires a shorter time than making a phone call and may be done at any time of day. Yet personal privacy and reimbursement remain difficulties in the majority of email settings, and patients with computer access are unlikely to be from the marginalized community with little income. Asynchronous telemedicine employs store-and-forward systems that may send photos, recordings, audio clips, health records, and large data sets from one location to another.

ii. **Synchronous/Interactive telehealth:** Interactive telehealth permits healthcare professionals to communicate with their patients in real time from faraway places via conferencing and technology-enabled healthcare equipment. Medical exams are carried out using transportable video cameras fitted with specific lenses that capture pictures along with other data that is subsequently sent to a large medical centre for diagnosis and treatment. Technology advancements allow healthcare professionals to access digital health records while still interacting with the person in need. Such devices also save data from the present session for later review or to track patient improvements. Interactive systems can be constructed on an individual computer that operates health record software and includes videoconferencing and data sharing capabilities. Since there are numerous uses for telemedicine technologies in medical practices, the following types of medical settings frequently use one or more of them: home health care facilities, rural medical facilities, community hospitals, and scattered medical professionals conducting clinical studies.

9.2.3 Various types of telehealth services

Telehealth refers to the adoption of digital tools to provide medical facilities as well as data remotely. There are several sorts of telehealth facilities, each having a unique purpose in healthcare delivery. Some common types of telehealth include:

i. **Live Video Conferencing (Synchronous Telehealth):** This involves immediate interaction between health care professionals and their patients through video conferences. It supports virtual consultations, follow-ups, and discussions about symptoms or treatment plans.

ii. **Store-and-Forward (Asynchronous Telehealth):** In this kind of telehealth, medical information, like photographs, videos, or patient data, is recorded and then sent to a healthcare professional for further evaluation. This method is often used for diagnostics, as the healthcare provider can assess the information conveniently.

iii. **Remote Patient Monitoring (RPM):** RPM is an area of telehealth care that involves people using mobile healthcare equipment and modern technology to obtain patient-generated health data (PGHD) and communicate it to medical treatment providers. RPM is the use of technology for gathering and exchanging patient health data from a distance. This can contain indicators of wellness, exercise levels, and other important health measures. It is especially useful for controlling chronic illnesses and providing immediate assistance. RPM is widely used to assist patients who require long-term, post-discharge, or elderly care.

iv. **Mobile Health (mHealth):** This term refers to the usage of mobile technology, such as smartphones and tablets, to assist with healthcare delivery, as well as wearable technology, such as smart watches, to offer health-related data about the wearer. It may include mobile apps for health monitoring, prescription reminders, and virtual consultations. mHealth refers to patients using mobile communication technology, such as personal digital assistants.

v. **Telemedicine Platforms:** Telemedicine is the technology of providing health consultation and counselling from a remote location through ICT. These are comprehensive online platforms that provide a range of telehealth services, including video consultations, secure messaging, and the sharing of health records. These platforms may integrate various telehealth technologies to offer a complete healthcare solution. Like a telephonic interview, telemedicine act as helping people to take health advice on call. It is more helpful to OPD patients during COVID-19 pandemic. The Indian Space Research Organization (ISRO), the Department of Information and Technology (DIT), the Ministry of Communications and Information Technology, the Ministry of Health and Family Welfare, the Ministry of External Affairs, and the Indian state governments are all working to establish telemedicine networks across the country.

vi. **Interactive Voice Response (IVR):** IVR systems use pre-recorded voice messages and keypad inputs to provide information and collect patient data. This can be used for appointment reminders, medication adherence, and basic health information.

vii. **Telepharmacy:** Telepharmacy allows pharmacists to provide pharmaceutical care remotely. Patients can consult with pharmacists via phone or video to discuss medications, side effects, and other concerns. Telepharmacy has been shown to potentially be used as an alternative method of dealing with drug.

viii. **Teledentistry:** This involves using telehealth technologies to provide dental care remotely. Dentists can conduct virtual consultations, assess oral health, and provide guidance on dental issues.

ix. **Telepsychiatry:** Mental health services can be delivered remotely through telepsychiatry, allowing human beings to seek therapy, counselling, or psychiatric consultations from the ease of their own residences.

x. **Teleradiology:** This involves sending medical photographs, which include X-rays, CT scans, or MRIs, to radiologists for examination and diagnosis from a remote place.

xi. The adoption of telehealth is growing significantly, especially with technological developments and the growing demand for remote medical services, making it a versatile and valuable tool in modern healthcare delivery.

9.3 E-learning

The core element of e-learning refers to scenarios in which both instructor and learner reside in different locations and depend upon electronic equipment for providing instruction-learning refers to learning that can be obtained electronically, either in a part or completely, using a web browser to access the internet, an intranet, or other forms of multimedia. E-learning includes the application of digital content and ICTs (information and communication technology) in education. This encompasses a wide range of media that show written content, audio, visuals, videos, and watching video; technological applications and procedures such as audio or video tape, satellite TV, a compact disc, and computer-based education; and local intranet/extranet or online education. Online learning is also known as virtual, distributed, network, and online learning.

According to UNESCO (2020), the closure of colleges and schools has numerous negative consequences for students, for example, interrupted learning, leading to students and youth being denied opportunities for development and progress. The study by Robles, Guerrero, Llinas, and Montero (2019) found that using mobile devices in the classroom is an interesting characteristic of how the educational system is growing. Furthermore, considering applications can assist both teachers and students in facilitating the teaching and learning process outside of the classroom. Webinar (Web+Seminar) and Webcon (Web+Conference) have recently become more famous because of the lockdown due to the coronavirus. NIC VC, Vidyo Desktop, Zoom, WebEx, Teams, Jitsi, and Google Meet are a portion of the video conference (VC) tools that were broadly utilized in Government, private segment, and educational institutions during the lockdown and post-lockdown period for controlling COVID-19 pandemic in India. Zoom is crucial to WFH and distance learning classrooms in India (Bajaj, 2020). Zoom is the most utilized software application for group discussion; approximately 100% of students utilize Google as a web engine. Educational learning systems find support from ICT, such as online tests, Interactions, evaluation tests, schoolwork, activities, and student communication (Kulkarni and More, 2020). Online education encourages teachers to become "Guides on the Side" instead of "Sages on the Stage" (Prensky, 2008). Encouraging professional development (PD) among teachers is pivotal to the nature of education. This works well during the pandemic when everyone who teaches suddenly needs to be on the internet. Content knowledge (CK) is also harder to address when the intended interest group comprises all higher educational institution teachers in the Netherlands, for example, nearly 50,000 educators (Schildkamp *et al.*, 2020). Most youngsters will have extraordinary trouble getting to online learning, the effect of which will likely be of restricted incentive for a large portion of them. The 'best' students in off-line learning situations will in general perform better than their friends in internet learning conditions too. Web-based learning for young students is more troublesome, whereas older students do better. Making information accessible on an enormous assortment of electronic devices and compatible with mobile devices is basic (World Bank, 2020). The coronavirus pandemic causes numerous social changes in public behaviour. Learners that need to learn from home: Ruangguru is the head of Indonesia's computerized learning organization (Fattah and Sujono, 2020).

9.3.1 Classification of E-learning

Based on the delivery mechanism, there are three basic types of e-learning as follows:

i. **Synchronous Education Delivery:** Students and teachers interact in the real predetermined time, and generally, it is the teacher-led environment (instant communication and clarification), for example, text chat, whiteboards, etc. In synchronous learning, employing satellite downlinks, video conferencing, or internet access, students and the teacher have a session that resembles conventional education through technological devices. Satellite downlinks are similar to television shows that frequently transmit information and training sessions covering a variety of subjects. The instructor is available by telephone for questions, and classes can be recorded for future viewing.

ii. **Asynchronous Education Delivery:** It is completely student-guided learning. Content is located at a centrally accessible location like website round the clock with conditional or unconditional access. Learners who participate in asynchronous remote learning have access to studying independently resources and modules at their own pace, mostly through the Internet. The capability of learners to access the resources and complete the activities on their own schedule is the most significant advantage of the asynchronous mode. This gives students a lot of flexibility to accommodate the course into your hectic schedule. Early courses that utilized the internet were substantially like correspondence courses in that they posted materials online instead of through regular mail. Courses designs are evolving and have ranged from simple web pages to elaborate presentations with multimedia. Interaction between students and teachers is increased when real-time and posted Internet bulletin boards are available for discussions.

iii. **Blended learning:** Blended learning is also referred to as hybrid learning, technology-mediated education, or mixed-mode instruction. In blended learning, online and offline (conventional, face-to-face) learning are combined so that each enhances the other. It can integrate human intervention—whether nearby or remote—with software-based resources. It can also integrate software-driven resources with any other type of educational resource, including books, tapes, TV, radio, and computer-mediated communication (like email or chat).

9.3.2 Merit of e-learning

E-learning has become an essential component of modern educational and training programs, offering a range of benefits that cater to the diverse needs of learners and institutions alike. Educational possibilities have expanded tremendously for people who cannot enrol in conventional classes due to time or location restrictions. The merits of e-learning are as:

i. Connecting to the unable-to-reach students in remote areas.
ii. Facilitating education for students from rural areas with multilingual backgrounds in the medium of their own choosing.
iii. Countering the insufficient teacher-student ratio.
iv. E-learning enables students to access educational materials at their preferred pace and convenience, allowing for more flexible study schedules.
v. Compensating for a lack of excellent educators through offering everyone in need with the opportunity to use educational materials created by outstanding educators.
vi. Ensure a unified curriculum across the country, provide equal chances for all students to access content for increased learning, and improve educational quality.
vii. This makes the learning process more participatory and entertaining.
viii. It saves both resources and time.
ix. E-learning significantly reduces the need for printed materials, contributing to environmental sustainability.
x. Learning becomes a continuous effort.
xi. The instructional method is more reliable, which benefits slow learners.
xii. Gives virtual support to learners with any type of obstacle.
xiii. E-learning platforms can provide instant feedback on assessments.

9.3.3 Demerit of E-learning

A study carried out in Kenya identified three major challenges to e-learning: an insufficient ICT foundation, a lack of specialized skills, and financial limitations (Tarus *et al.*, 2015; Almaiah *et al.*, 2020). The following are demerit of e-learning:

i. **Health-related concerns:** E-learning may lead to physical problems, such as eyestrain, poor posture, and other problems, due to the use of computers and devices.
ii. **Cyber wandering:** Because of lack of guidance, often a learner may get distracted to do activities other than what he or she is supposed to do.
iii. **Need for technical expertise:** Sometimes, technical expertise is needed to access and manage the content in e-learning. This could be due to technological compatibility issues between hardware and software.
iv. **Need to change 'constantly':** Since technology changes so rapidly, a learner who practices e-learning must always stay updated with technology.

9.4 Conclusion

The exploration of telehealth and e-learning in this chapter underscores their pivotal role in transforming access to healthcare and education across the globe. As we have seen, these innovations offer unprecedented convenience and extended reach, providing services and knowledge to previously underserved or inaccessible populations. However, as with any technological advancement, they introduce a set of challenges that must be navigated carefully to maximize their potential benefits while minimizing risks. Telehealth, while breaking geographical barriers and making healthcare more accessible, poses significant challenges regarding data security and the need for regulatory adjustments. Protecting patient data is paramount, requiring robust encryption methods, secure access controls, and continuous monitoring to safeguard information integrity. Additionally, the shift towards a more digitally interconnected healthcare environment necessitates a reevaluation of clinical licensure and regulatory frameworks to adapt to the global nature of telehealth. Similarly, e-learning democratizes education, enabling learners from any location to access high-quality educational resources and interactive learning environments. However, this learning mode demands substantial enhancements in digital literacy and infrastructure, especially in developing regions. Ensuring equitable access to technology and addressing diverse educational needs are essential to prevent the widening of the digital divide.

Both telehealth and e-learning exemplify the dynamic intersection of technology and human need. Their continued evolution will likely depend on a multidisciplinary approach involving technological innovation, policy reform, and ongoing research to address the complexities of delivering services remotely. By focusing on inclusivity and security, we can harness the full potential of these platforms to foster a more informed and healthier global population. In the interconnected world of the 21st century, telehealth and e-learning are not merely optional; they are essential components in the progression towards more accessible, effective, and equitable healthcare and education systems. As we move forward, it is crucial that all stakeholders governments, institutions, technology providers, and the community work collaboratively to overcome the hurdles and shape a future where technology serves as a bridge, not a barrier, to health and knowledge.

References

Advanced Market Research for Telemedicine Health Care.www.feed-back.com/usstethoscopes.html

Bajaj, K. (2020). Post Coronavirus World - Technology Imperatives: Make in India for self-reliance. Retrieve from

Bajpai, N. and Wadhwa, M. (2019). Accessing specialist services via Telemedicine in India. ICT India Working Paper #21, Center for Sustainable development, Earth Institute, Columbia University.

Borowitz, S.M. and Wyatt, J.C. (1998). The origin, content, and workload of e-mail consultations. JAMA; 280:1321-1324.

Brwon, N. (1996). What is telemedicine? Association of Telehealth Service Providers, 1996. http://atsp.org.

Cermack, M. (2006). Monitoring and telemedicine support in remote environments and in human space flight. British Journal of Anaesthesia, 97(1): 107-114.

Codish, S., Novack, L., Dreiher, J., Barski, L., Jotkowitz, A., Zeller, L. and Novack, V. (2014). Impact of Mass Media on Public Behavior and Physicians: An Ecological Study of the H1N1 Influenza Pandemic. Infection Control & Hospital Epidemiology, 35(6): 709-716.

Collinson and Heffernan (2014). Modelling the effects of media during an influenza epidemic. BMC Public Health, 14: 376.

Dat, T.V., Tu, V.L., Quan, N.K., Minh, N.H., Trung, T.D., Le, T.N., Phuc-Vinh, D., Trinh, D.T., Pham Dinh. L., Nguyen-Thi, H.Y., Huy, N.T. (2023). Telepharmacy: A Systematic Review of Field Application, Benefits, Limitations, and Applicability During the COVID-19 Pandemic. Telemed J E Health. 29(2): 209-221.

Fattah, R.A. and Sujono, F.K. (2020). Social presence of Ruangguru in social media during COVID-19 pandemic. Jurnal The Messenger, 12(2): 180-191.

Harun, M.H. (2001). Integrating e-Learning into the workplace.The internet and higher education 4(3-4): 301-310.

HRSA Telehealth Programs Web Page.https://www.hrsa.gov/rural-health/telehealth#:~:text=Telehealth%20is%20defined%20as%20the,public%20health%2C%20and%20health%20administration.

Iyengar, S. (2020). Mobile health (mHealth). Fundamentals of Telemedicine and Telehealth. Academic Press, 277-294.

Kadir, M.A. (2020). Role of Telemedicine in Healthcare during the COVID-19 Pandemic in the Developing Countries. Telehealth and Medicine Today.Pages 1-5.

Kulkarni, M.A. and More, H.V. (2020). Role of Information and Communication Tools (ICT) in Higher Educational Institutional System during Covid-19 Lockdown Period.International Journal of Information Technology, 6(4): 12-16.

Kumar, S., Sharma, B. and Singh, V. (2020). Modelling the role of media induced fear conditioning in mitigating post-lockdown COVID-19 pandemic: perspectives on India. arXiv:2004.13777

Leask, J., Hooker, C. and King, C. (2010). Media coverage of health issues and how to work more effectively with journalists: a qualitative study. BMC Public Health, 10:535. Retrieved from http://www.biomedcentral.com/1471-2458/10/535

Mejia, C.R., Ticona, D., Rodriguez-Alarcon, J.F., Campos-Urbina, A.M., Catay-Medina, J.B., Porta-Quinto, T., Garayar-Peceros, H., Ignacio-Quinte, C., Carranza Esteban, R.F., Ruiz Mamani, P.G. and Tovani-Palone, M.R. (2020). The Media and their Informative Role in the Face of the Coronavirus Disease 2019 (COVID-19): Validation of Fear Perception and Magnitude of the Issue (MED-COVID-19). Electronic Journal of General Medicine 2020; 17(6):em239.

Prensky, M. (2008). The Role of Technology in teaching and the classroom. Educational Technology, Nov-Dec, 2008.

Ringel, M. (2000). Physician Shares Views from the Trenches. Minneapolis, MN, Association of Telehealth Service Providers Fourth Annual Conference.

Robles, H.R., Guerrero, J., Llinas, H. and Montero, P. (2019). Online teacher- students interactions using WhatsApp in a law course. Journal of Information Technology Education: Research, 18: 231-252.

Schildkamp, K., Wopereis, I., Jong, M.K., Peet, A. and Hoetjes, I. (2020). Building blocks of instructor professional development for innovative ICT use during a pandemic.Journal of Professional Capital and Community.Retrieve fromhttps://doi.org/10.1108/JPCC-06-2020-0034

Shakeel, S.; Nesar, S., Rahim, N., Iffat, W., Ahmed, H.F., Rizvi, M. and Jamshed, F. (2017). Utilization and Impact of Electronic and Print Media on the Patients' Health Status: Physicians' Perspectives. Journal of Pharmacy and Bioallied Science. 9(4): 266-271.

Tarus, J. K., Gichoya, D. and Muumbo, A. (2015). Challenges of implementing e-learning in Kenya: A case of Kenyan public universities. The International Review of Research in Open and Distance Learning, 16(1): 120-141.

Telemedicine Information Exchange. National Library of Medicine.http://tie.telemed.org.

Tuckson, R.V., Edmunds, M. and Hodgkins, M.L. (2017). Telehealth.The New England Journal of Medicine. 377(16): 1585-1592.

University of Virginia Health System Office of Telemedicine. www.telemed.virginia.edu

World Bank (2020). Rapid Response Briefing Note : Remote Learning and COVID-19 Outbreak (English). Washington, D.C.: World Bank Group. Retrieved from

World Health Organization, ed. (2010). Telemedicine: Opportunities and Developments in Member States: Report on the Second Global Survey on E-Health. Global Observatory for E-Health Series 2. Geneva, Switzerland: World Health Organization.

10

E-Commerce and Entrepreneurship Bridging Markets and Creating Livelihoods

Kumari Asha[1] and Rohit[2]

[1]G.L.M. Agriculture College, Kishangarh bas, Alwar (Affiliated to S.K.N. Agriculture University, Jobner, Jaipur), Rajasthan, India

[2]College of Agriculture, Lords University, Chikani, Alwar, Rajasthan, India

Abstract

This chapter explores the convergence of e-commerce and entrepreneurship, highlighting its potential to bridge markets and create new livelihood opportunities. It discusses the various approaches to e-commerce (B2B, B2C, C2C, etc.) and emphasizes its advantages, including wider reach, reduced costs, and increased customer choice. The chapter also acknowledges e-commerce's challenges, particularly in rural areas, where issues like digital literacy and infrastructure gaps hinder wider adoption. Success stories of Indian e-commerce giants like Snapdeal and Flipkart are presented as case studies, showcasing their strategies and contributions to the sector's growth. Finally, the chapter explores initiatives to bridge the digital divide and empower rural micro-entrepreneurs to participate in the e-commerce revolution.

Keywords: *E-commerce, Entrepreneurship, Digital Transformation, Rural Development, Billion-Dollar Businesses.*

10.1 Introduction

The incredible technology usage in e-commerce over the last several years has resulted in the fast growth of e-business enterprises. Technological innovation has created new opportunities for online businesses looking to provide value-added services. Technology adoption, enhanced customer support, and creative product development have given businesses the necessary competitiveness. Consumers are more demanding than ever and more aware of their rights. E-commerce businesses are pressured to develop and provide value for their customers in novel ways.

E-commerce encompasses all forms of business done through computer networks. Global e-commerce has undergone a structural transformation due to the progress of information and technology. A lot of facets of our lives have been revolutionized by digital technology. The competitive landscape and the ever-evolving technologies and lifestyles have transformed e-business. Competition will be fierce given the opening of the Indian economy and the government's ongoing push for privatization. Computer networks are now a necessary component of daily life due to recent advancements in telecommunications and computer technology. The number of businesses that enable online transactions is growing every day in the modern day. Online shopping offers customers several advantages, including faster delivery, more options, reduced prices, and more time efficiency. With just a mouse click, anyone may purchase items without leaving their home or business.

10.2 Meaning and Concept of E-Commerce

Technology is transforming the entrepreneurial environment. These days, electronic commerce is the buzzword in business. E-commerce is the practice of conducting business online or electronically to increase operational efficiency. This process strengthens the value offered to clients and gives a firm a competitive edge over rivals. Products and services are purchased and sold in business-to-business, business-to-consumer, and consumer-to-consumer transactions. E-commerce stokes the flames of invention if entrepreneurship is innovation.

E-commerce is a modern approach that meets entrepreneurs' demands. Service delivery is expedited, and costs are decreased. While e-commerce offers entrepreneurs the opportunity to start a new business, those who currently have one should work to improve their online visibility. Having a website might benefit marketing and sales campaigns. Entrepreneurs should utilize these venues to understand how the Internet can transform and grow their businesses. For example, a local grocery seller who may largely serve the neighborhood might see increased customers around the city by developing a website.

10.3 Approaches of E-Commerce

i. **Business-to-business E-Commerce:** B2B e-commerce is the term used to describe e-commerce between businesses. Business-to-business (B2B) partnerships are the focus of e-commerce. It considers supply chain management, electronic procurement, and online transaction negotiation for the businesses. E-markets and e-frastructure are the two primary parts of B2B e-commerce. The logistics-transportation, storage and distribution, and application service providers make up the majority of the B2B e-frastructure. E-markets are online marketplaces where merchants and buyers communicate and transact business. Cisco, Dell, IBM, and Hewlett-Packard (HP) are a few B2B examples. In addition to reducing transaction costs, business-to-business e-commerce also saves time and effort.

ii. **Business-to-Consumer (B2C)** E-Commerce: This type of e-commerce is the second most popular and oldest kind. Businesses use internet technology to introduce goods and services to customers. B2C e-commerce entails consumers obtaining data and making purchases of tangible items and information products via an electronic network. Online retailers like Firstcry.com, Flipkart.com, and Snapdeal.com are instances of business-to-consumer e-commerce. One of the key advantages of business-to-consumer e-commerce is that it allows business owners to grow their customer base while incurring the fewest expenses possible. Through business-to-consumer e-commerce, buyers may locate the best deal on a good or service.

Fig. 10.1: Approaches of E-Commerce

iii. **Consumer-to-consumer E-Commerce**: The Consumer-to-Consumer (C2C) approach to e-commerce focuses on how people use it to trade goods and information. The emergence of online auctions and electronic marketplaces, especially in vertical sectors where businesses may bid on products from various providers, characterizes(C2C) e-commerce.

iv. **Government to Business (B2G):** E-commerce: Business-to-government (B2G) e-commerce is the term used to describe online transactions between businesses and the public sector. The requirement for businesses to sell products or services to governments or government agencies is the focus of business-to-government e-commerce. One example of such an activity is providing goods and services to hospitals, schools, the police, and the army.

v. **Mobile Commerce:** Mobile commerce, abbreviated as M-commerce, refers to the exchange of goods and services using wireless technology, primarily through mobile devices. It is recognized for its swiftness and enhanced security compared to traditional digital commerce methods. M-commerce encompasses various financial activities such as mobile banking, brokerage services, telecommunications, bill payments, and more.

10.3.1 The Economics of Attention: Social Media and E- E-Commerce Potential

Since the emergence of social media behemoths like Facebook (2005), YouTube (2005), and Twitter (2007), millions of individuals worldwide have been able to interact more than ever before and perhaps more importantly create and share original material quickly and easily. The innumerable videos, comments, news stories, and status updates continuously uploaded and debated on social media sites like Facebook, Digg, and Twitter are evidence of the widespread engagement in these networks. The numbers these websites brag about and the worldwide demographics that regularly use these services demonstrate the impact of their growth. The capacity of people to communicate information freely and have rapid access to knowledge that would have been difficult or impossible to discover in the past defines the Information Age, which we are currently living in. As a result, the widespread creation and consumption of information has led to the emergence of a fiercely competitive online marketplace where various forms of material compete for users' limited attention. Despite the seemingly disorganized way these exchanges occur, some subjects get an excessive amount of attention, rising to the top of the popularity scale. Due to their prominence, these hot subjects help raise everyone's knowledge of current trends. Through their visibility, these popular topics contribute to the collective awareness of what is trending and at times can also affect the public agenda of the community.

Fig. 10.2: Benefits of E-Commerce

10.4 Advantages of E-Commerce

i. All geographical restrictions have been eliminated by e-commerce. An e-commerce website opens up the globe to entrepreneurs.

ii. By using several search engines to find new clients, the e-entrepreneur might increase the size of its clientele.

iii. The reduced cost is one of the most particular benefits of e-commerce. An e-commerce business owner does not need to have costly retail showrooms or enormous stockpiles. Their sales and marketing staff is far less than those of actual retail establishments. Discounts are one way that businesses may give their consumers a piece of the savings from these reduced costs.

iv. Online markets offer consumers an extensive array of choices. Through ecommerce platforms, shoppers can swiftly refine their searches using user-friendly navigation or a search function. Some sites even store customer preferences and shopping lists to streamline future purchases.

v. Ecommerce fosters price comparison, with numerous online tools enabling customers to browse various merchants and locate the most competitive prices. With information readily accessible, consumers can compare products, features, prices, and read reviews before making a selection.

vi. Consumers benefit from broader access to assistance and guidance, whether from experts or peers. Unlike physical stores, ecommerce websites can provide abundant supplementary information to customers. Additionally, consumers appreciate prompt service and delivery of their purchases, with orders conveniently shipped directly to their homes. By reducing expenses, ecommerce allows companies, particularly smaller ones, to globally disseminate information about their products and services to potential customers.

10.5 Disadvantages of E-Commerce

Though e-commerce is widely accepted all around the world but still it has certain disadvantages which are discussed as under:

i. Calculating return on investment is challenging.

ii. Many businesses have struggled to find and keep workers with the expertise in business process management and technology design required to have a strong online presence.

iii. The challenge of incorporating transaction processing software and databases that are currently in use and intended for traditional commerce with the software that facilitates e-commerce.

iv. Conducting e-commerce presents cultural and legal challenges for many firms.

10.6 Difference between Traditional Commerce and E-Commerce

A growing number of conventional small businesses are viewing e-commerce as an alluring and lucrative sales channel as a result of the advancements in information technology. E-commerce and conventional commerce, however, differ from one another for a variety of reasons. The following is a list of a few of the causes of the differences:

i. Customer Interaction: Traditional commerce relies on face-to-face interaction, allowing customers to engage with sellers, ask questions, and potentially lead to upselling. In contrast, e-commerce lacks this direct engagement unless features like live chats are integrated.

ii. Cost Savings: E-commerce eliminates the need for physical store setups in prime locations, reducing expenses like rent and electricity. Setting up an online store is significantly cheaper, enabling small business owners with limited capital to start their ventures.

iii. Global Reach: Unlike traditional commerce, online shopping transcends geographical boundaries, allowing businesses to reach customers worldwide. Online marketing further enhances sales potential, with no limitations on client base size.

iv. Product Returns: While traditional commerce allows customers to physically inspect products before purchase, online businesses face higher return rates as customers can order and try items at home. This absence of direct interaction may lead to increased returns.

v. **Credit Card Fraud:** E-commerce is vulnerable to credit card fraud, posing risks of identity theft during transactions. In contrast, traditional commerce offers more security through face-to-face interactions and verification processes like asking for a photographic ID.

Transitioning to online sales requires adapting to new customer interactions, marketing strategies, and order fulfillment methods. Despite challenges, the benefits include cost efficiency, global reach, and 24/7 business operations, allowing focus on product improvement and customer experience enhancement. Some products, like jewelry, thrive in online sales, while an online presence can also complement traditional commerce by attracting customers and showcasing products.

10.7 Registration and Licenses Required for Starting an E-commerce Business

10.7.1 Registration

The first thing to be aware of is that an internet firm that want to take payments online must be registered. They require this in order to set up a secure payment gateway on the website. There are three methods available for registering a business. Registering as a Sole Proprietor is the simplest and most practical method to start your own business. A sole proprietorship has several advantages, including fewer paperwork and legal compliance. Additionally, Indian law recognizes the owner and their firm as one and the same entity, which allows you to operate your business out of your house (unlike Pvt. Ltd. and public businesses). Individual tax rates are used to pay taxes; corporation tax rates are not used. All you need is a bank account at a nearby bank branch in the name of your business; there are no extra fees. You can file taxes using your personal PAN card. Because there is minimal turnover in the early stages of an e-commerce firm, the sole proprietorship is the most common structure for these companies. Depending on the size of your company, the second option for registration is to choose between a Limited Liability Partnership (LLP), a Public Company, or a Pvt Ltd. Here, the first thing you need do is open a current account in your company's name.

Fig. 10.3: The Process of E-Commerce

10.7.2 Taxation

If you operate as a sole proprietor, you can use your own PAN card for tax reasons. Income tax returns must be filed in your name. Since a sole proprietorship pays individual rates, there is no corporation tax to be paid. Only services rendered are subject to service tax. Sales tax, excise tax, and VAT are all applicable to the sale of products. Only if your company's annual revenue exceeds 10 lacs is it required to pay these taxes.

10.7.3 LNavigating Online Business Licenses: A State-by-State Adventure

Running an online portal requires specific licenses depending on your offerings. Here's a breakdown to get you started:

i. **Physical Presence and Payments:** If you plan to have a physical store or utilize a payment gateway, you'll likely need a Shops and Establishment Act registration. This also applies if you hire employees.

ii. **Taxation Thresholds:** In India, online sales come with tax implications, but only once your annual turnover surpasses certain limits. Here's a basic overview:

a) **CST/VAT:** This applies to the sale of goods online and kicks in once your yearly revenue exceeds ₹5 lakhs.

b) **Service Tax:** Levies a tax on online service offerings, but only after your annual turnover crosses ₹10 lakhs.

c) **Employee Management:** Businesses with even one employee must register for Professional Tax.

d) **Remember, tax and registration rules vary by state.** Always consult your local authorities for the most up-to-date information specific to your location.

10.7.4 E-commerce Platforms: Empowering Rural Micro-Entre-preneurs in India

This section explores how e-commerce platforms in India facilitate micro-entrepreneurs' participation, particularly those in rural areas.

10.8 Decentralized Ownership for Local Benefit

Inspired by a model for local ownership of cellular networks in Indonesia ([Heimerl *et al.*, 2013]), there's potential to adapt a similar approach for e-commerce in India. This "bottom-up" model would empower local entrepreneurs to own and operate e-commerce platforms catering to their communities. This reduces operational costs, creates local opportunities, and fosters a sense of autonomy (as discussed by [Sen, 2001]).

10.8.1 Existing Platforms and Government Initiatives

The Indian government actively encourages rural micro-entrepreneurs to leverage digital channels. This can be achieved through established e-commerce platforms with robust distribution networks, dedicated government portals, or even through independent social media storefronts.

For instance, Uttar Pradesh's "One District One Product" (ODOP) program, backed by the Ministry of MSME, promotes indigenous crafts from each district. Amazon India has partnered with the state government to support these micro-enterprises by providing training, account management, marketing tools, and access to their storage and delivery infrastructure ([Naik, 2019]). This aligns with the "Atmanirbhar Bharat" initiative emphasizing domestic production and distribution. Similar collaborations exist between Flipkart and the state of Karnataka through the "Samarth" program, which offers incubation support and assistance with onboarding, marketing, and logistics ([Economic Times, 2020]).

The government is also planning a dedicated e-commerce portal for MSMEs called "Bharat Craft" in collaboration with the State Bank of India (SBI). This initiative aims to bypass middlemen, mirroring the model of China's e-commerce giant, Alibaba ([Economic Times, 2020]).

10.8.2 Social Media as a Marketplace

Beyond formal e-commerce platforms, many rural sellers are finding success by utilizing social media for direct marketing. As highlighted in Jack *et al.*'s study ([Jack *et al.*, 2017]), these online ecosystems thrive on adapting to local contexts. Platforms like "Dukaan" enable local shops to create online storefronts and manage product listings, pricing, and delivery through popular social media platforms like WhatsApp, Facebook, and Instagram. This empowers micro-entrepreneurs to reach a wider audience directly.

10.8.3 Independent Initiatives

The Khadi and Village Industries Commission (KVIC) exemplifies an independent approach. With onboarding hurdles with major e-commerce platforms, KVIC launched its e-commerce portal for selling khadi products (hand-woven cloth) including masks, fabrics, food items, and more. This initiative proved highly successful, prompting interest from other e-commerce giants. However, the government's trademark protection for "Khadi" ensures KVIC remains the sole authorized seller ([KVIC, 2020]).

10.9 E-Commerce Struggles in Rural India: A Bridge Too Far?

Despite the growing popularity of e-commerce in India, a significant gap exists between urban and rural adoption. This disparity stems from several key challenges:

i. **Digital Literacy Gap:** Studies by The Boston Consulting Group (20??) highlight the user interface struggles faced by rural consumers. Low literacy rates and limited technical expertise, as identified by Anooja (2015) and Goswami (2016), restrict effective e-commerce utilization. This lack of digital fluency is further exacerbated by the urban-rural divide in literacy rates, as highlighted in the Proceedings of the 1st Virtual Conference on Implications of Information and Digital Technologies for Development (2021).

ii. **Information Accessibility:** UNESCO's definition of literacy emphasizes its use as a communication tool. Parikh *et al.* (2003) point out India's high illiteracy rate (43%), with some northern states reaching a staggering 70-80%. This lack of basic literacy hinders the ability to leverage available technology. Furthermore, Bakshi (2019) emphasizes the additional challenge of inadequate customer support literacy in rural areas.

iii. **Trust and Security Concerns:** Rural microentrepreneurs fear online fraud, which discourages them from embracing e-commerce platforms.

Goswami and Dutta (2016) attribute this apprehension to a combination of limited education and a lack of financial and social support systems.

iv. **Digital Knowledge Management:** While mobile phone penetration is increasing in rural areas, effective use of digital information remains a hurdle. This knowledge gap hinders the full potential of e-commerce adoption.

v. **Geographical Challenges:** Poor rural infrastructure creates significant logistical roadblocks. The lack of efficient transportation networks makes shipping goods to and from remote areas a major challenge, as highlighted by Khatri (2019). While India Post has seen a resurgence by collaborating with e-commerce giants (Karnik, 2016), its reach might not be sufficient for widespread rural coverage.

Bridging this digital divide requires a multi-pronged approach that addresses literacy gaps, enhances information accessibility, builds trust in online transactions, and invests in rural infrastructure development. Only then can the full potential of e-commerce be unlocked for all of India's citizens.

10.10 E-commerce Entrepreneurs: Building Billion Dollar Businesses

10.10.1 Snapdeal

From Deals to Domination: Snapdeal's story began in 2010 with a focus on offline couponing. Founders Kunal Bahl and Rohit Bansal saw success early on, selling 15,000 coupons in just three months. However, their ambition led them to seek venture capital funding. While their initial meeting with Vani Kola wasn't fruitful, her firm ultimately invested in Snapdeal, recognizing its potential.

Snapdeal's first foray into e-commerce proved challenging. They learned valuable lessons from early mistakes, fostering a culture of hard work and dedication to customer satisfaction. A pivotal moment arrived in 2011, inspired by Alibaba's success. Snapdeal boldly decided to pivot from deals to an online marketplace, shutting down their established business.

This risky move surprised investors, as eBay was the sole marketplace in India at the time. Yet, Rohit Bahl's determination secured board approval, paving the way for Snapdeal's present form. Snapdeal boasts a billion-dollar valuation, a testament to the founders' vision.

Snapdeal's success is fueled by several factors. Over 50,000 sellers offer a vast selection of 5 million products. Strategic partnerships, like the investment from eBay, brought valuable experience. With a focus on scalability and speed, Snapdeal has become one of India's fastest-growing e-commerce companies.

The youthful energy of their workforce, with an average age of 25, embodies their core values: innovation, change, openness, honesty, and ownership. These values propel Snapdeal towards continued success. While great ideas are crucial, Snapdeal's true strength lies in the confident execution, marketing and unwavering pursuit of becoming the top B2C marketplace in India.

10.10.2 Flipkart

From Books to Billion: Flipkart's story, like many successful ventures, began with a groundbreaking idea. Founded in 2007 by Sachin and Binny Bansal, former Amazon employees, Flipkart entered the nascent Indian e-commerce market. This was a daring move, considering the diverse consumer preferences and traditional shopping habits in India. However, Flipkart's success has been instrumental in transforming e-commerce into one of India's fastest-growing sectors. Flipkart started by selling books, quickly expanding its product range. They pioneered innovative features for the Indian market, such as the now-ubiquitous "Cash on Delivery" option. Early funding from venture capitalists fueled their growth, with Flipkart consistently exceeding expectations. Their sales grew from 40 million rupees in 2008-09 to 200 million rupees the following year, aiming to reach the billion-dollar mark by 2015.

Two major challenges confronted e-commerce companies in Flipkart's early days: online payment gateways and supply chain management. Flipkart addressed the payment hurdle by offering cash-on-delivery and card-on-delivery options alongside online payment methods. They established their robust supply chain system, recognizing the importance of timely deliveries. Flipkart's remarkable journey from a small book retailer to India's e-commerce leader serves as an inspiration for a generation of aspiring entrepreneurs.

10.11 Conclusion

E-commerce is a contemporary method that caters to business owners' needs. Costs are reduced and service delivery is accelerated. Even while e-commerce is a chance for entrepreneurs to launch a new company, those who already have one should endeavors to strengthen their online presence, having an online presence may help with sales and marketing initiatives. Entrepreneurs should use these platforms to learn how the Internet may change and expand their firms. By creating a website, a neighborhood grocery seller, for instance, who could be primarily serving the neighborhood, might increase the number of people who shop with them throughout the entire city. E-commerce encompasses all forms of business done through computer networks. Global e-commerce has undergone a structural transformation due to the progress of information and technology. A lot of facets of our lives have been revolutionized by digital technology. The competitive landscape and the ever-evolving technologies and lifestyles have transformed e-commerce.

References

Abrar, P. (2020). Flipkart says 90% sellers back on platform, 125% rise in new MSMEs.

Business Standard. Retrieved November 30: 2020 from https://www.businessstandard.com/article/companies/flipkart-says-90-sellers-back-on-platform-125-rise-in-newmsmes-120062700492_1.html

Ahmed, S. I., Bidwell, N. J., Zade, H., Muralidhar, S. H., Dhareshwar, A., Karachiwala, B., & O'Neill, J. (2016, May). Peer-to-peer in the Workplace: A View from the Road. In Proceedings of the 2016 CHI Conference on Human Factors in Computing Systems (pp. 5063-5075).

Amazon Services (2020). Amazon Global Selling. Retrieved December 4, 2020 from https://services.amazon.in/services/amazonglobalselling/benefits.html#:~:text=By%20signing%20up%20to%20sell,%26%20Asia%2DPacifi c%20(Japan%2C

Anand, N. & Phartiyal, S. (2020, April 24). Facebook's big India deal portends a bruising time for SoftBank-backed Paytm. The Guardian. Retrieved from https://www.theguardian.pe.ca/business/reuters/facebooks-big-india-deal-portends-abruising-time-for-softbank-backed-paytm-441694/

Anooja A. (2015). Digital India with E-Commerce Revolution in Rural India: Transform India Digitally and Economically. Engineering International, 3(7).

Ara, I. (2020). As COVID-19 Impact Continues, UP Artisans Are Selling Their Goods Online. (November 2020). Retrieved December 7, 2020 from https://thewire.in/government/ascovid-19-impact-continues-up-artisans-are-selling-their-goods-online

Atmanirbhar Bharat Abhiyan (2020). Government of India. Retrieved December 1, 2020 from https://aatmanirbharbharat.mygov.in/

Bahree, M. (2018). E-Commerce Firms Will Find Their Next Set of Consumers In Rural India: Report. Forbes. Retrieved on December 3, 2020 from https://www.forbes.com/sites/meghabahree/2018/11/04/e-commerce-firms-will-find-theirnext-set-of-consumers-in-rural-india-report/?sh=3c93f0bb1721

Bakshi, A. (2019). Gearing Up Small Businesses for Marketing in the Digital Space: An Exploratory Study. Proceedings of 10th International Conference on Digital Strategies for Organizational Success, Available at SSRN: https://ssrn.com/abstract=3309290 or http://dx.doi.org/10.2139/ssrn.3309290

Barbier, E. B. (1987). The concept of sustainable economic development. Environmental conservation, 14(2): 101-110. Proceedings of the 1st Virtual Conference on Implications of Information and Digital Technologies for Development, 2021 247

Burrell J., and Toyama K. (2009). "What constitutes good ICTD research?." Information Technologies & International Development 5(3): 82.

Chakraborty, S. (2020, May 13). Covid-19 package: MSME definition changed, loans WORTH Rs 3-trn announced. Retrieved March 27, 2021, from https://www.businessstandard.com/article/economy-policy/corona-package-msme-definition-changed-loansworth-rs-3-lakh-cr-announced-120051301214_1.html

Digital India, Ministry of Electronics & Information Technology, Government of India. Retrieved from http://digitalindia.gov.in/ Digital MSME (2019).

Digital India (2017). Digital India, Ministry of Electronics & Information Technology, Government of India. Retrieved on August 11, 2017 from http://digitalindia.gov.in/ Digital India. (2017, August 11).

Economic Times (2020). Flipkart signs MoU with Karnataka govt to Promote local art, craft and handlooms (July 2020). November 15, 2020 from https://retail.economictimes.indiatimes.com/news/e-commerce/e-tailing/flipkart-signsmou-with-karnataka-govt-to-promote-local-art-craft-and-handlooms/76889875

Economic Times (2020). MSME sector created 11 crore jobs in India: Nitin Gadkari. Economic Times (October 2020). November 15, 2020 from https://economictimes.indiatimes.com/small-biz/sme-sector/msme-sector-created-11-crorejobs-in-india-nitin-gadkari/articleshow/78945901.cms?from=mdr

Economic Times (2020). SBI working on setting up e-commerce portal for MSMEs: Rajnish Kumar. Retrieved December 7, 2020 from https://economictimes.indiatimes.com/industry/banking/finance/banking/sbi-working-onsetting-up-e-commerce-portal-for-msmes-rajnish-kumar/articleshow/76663992.cms

Economic Times (2020, January 8). Paytm launches all in one QR for merchants. Economic Times. Retrieved from https://economictimes.indiatimes.com/smallbiz/startups/newsbuzz/paytm-launches-all-in-one-qr-formerchants/articleshow/73157049.cms

Economic Times (May 2020). Retrieved on November 10, 2020 from https://economictimes.indiatimes.com/small-biz/smesector/finance-minister-announces-revised-msme-definitions-no-different-betweenmanufacturing-and-service-enterprises/articleshow/75717694.cms?from=mdr

Economic Times (2017). India became world No.1 in mobile broadband after Reliance Jio Launch: Ambani - ET Telecom." The Economic Times Telecom. Retrieved on December 2, 2017 from telecom.economictimes.indiatimes.com/news/india-became-world-no-1-inmobile-broadband-after-reliance-jio-launch-ambani/61888647

Economic Times (2020). Finance Minister announces revised MSME definitions; no difference between manufacturing and service enterprises.

Heimerl, K., Hasan, S., Ali, K., Brewer, E., & Parikh, T. (2013, December). Local, sustainable, small-scale cellular networks. In Proceedings of the Sixth International Conference on Information and Communication Technologies and Development: Full Papers-Volume 1 (pp. 2-12)

https://www.mn.uio.no/ifi/english/research/groups/is/ifip-94/proceedings-virtual-conference-2021/proceedings-ifip-9.4-1st-virtual-conference-2021.pdf

Jack M., Chen J., and Steven J. Jackson (2017). "Infrastructure as creative action: Online buying, selling, and delivery in Phnom Penh." Proceedings of the 2017 CHI Conference on Human Factors in Computing Systems.

Khadi and Village Industries Commission. Ministry of Micro, Small and Medium Enterprises. Government of India. Retrieved December 2, 2020 from https://www.kviconline.gov.in/khadimask/prodouctsgrps.jsp

Naik, A. (2019). Amazon India Takes First Step to Attract MSMEs, SMB Sellers With Small Business Day. (December, 2019). Retrieved December 1, 2020 from https://inc42.com/buzz/amazon-india-takes-first-step-to-attract-msmes-smb-sellers-withsmall-business-day

Schumacher, E. F. (2011). Small is beautiful: A study of economics as if people mattered. Random HouseTimes of India (2020). Over 70,000 Indian exporters gear up for Black Friday, Cyber Monday sale through Amazon. Retrieved November 24, 2020 from https://timesofindia.indiatimes.com/business/india-business/over-70000-indian-exportersgear-up-for-black-friday-cyber-monday-sale-through-amazon/articleshow/79393988.cms

Sen, A. (2001). Development as freedom. OUP Oxford.

Sheetal, J. U., Purohit, D. N., &Anup, V. (2019). Increase in number of online services and payments through mobile applications post demonetization. Advances in Management, 12(1): 34-38.

Shirley, M. A. J. (2017). Impact of Demonetization in India. International Journal of Trend in Research and Development, 20-23.

Singh, P. & Singh, V. (2016). Impact of Demonetization on Indian Economy. Indian federation of united nation associations, New Delhi, India

Sircar, J. (2006). MSME Development Act, 2006 - Rules - 26 September 2006. Ministry of Micro, Small and Medium Enterprises. Government of India. Retrieved December 1, 2020 from https://msme.gov.in/sites/default/files/gazette%20596-E-English.pdf

Swadeshi movement. Britanicca Dictionary. Retrieved December 1, 2020 https://www.britannica.com/topic/swadeshi-movement

UN. (2021). Take action for the sustainable development goals - united nations sustainable development. Retrieved March 26, 2021, from https://www.un.org/sustainable development/ sustainable-development-goals

UNESCO. (2018). Defining Literacy. (October 2018). Retrieved October 20, 2018 from http://gaml.uis.unesco.org/wp-content/uploads/sites/2/2018/12/4.6.1_07_4.6-definingliteracy.pdf

Vachhatani, J. (2020). KVIC Forces Amazon, Flipkart & Other E-Commerce Giants To Take Down Khadi Brand Products. Republic World. Retrieved December 6, 2020 from https://www.republicworld.com/india-news/general-news/kvic-forces-amazon-flipkartand-other-e-commerce-giants-to-take-down-kh.html

Varshney, R. (2020). Amazon sellers can now register and manage services in Hindi. (June 2020). Retrieved December 1, 2020 from https://yourstory.com/2020/06/amazon-launches-sellerregistrations-account-hindi

Wasan, P. G., & Jain, N. (2017). Customizing content for rural mobile phones: a study to understand the user needs of rural India. Social Network Analysis and Mining, 7(1): 1-13.

11

Gendered Innovations: Designing Technology for Women by Women in Agriculture

Md. Mubeena[1], S. Kavitha[2], K. Balaji Naik[3], T. Lakshmi[1] and V. Sailaja[1]

[1]*Department of Agricultural Extension Education, S.V.Agricultural College Tirupati, India*

[2]*Department of Agricultural Extension Education, Professor Jayashankar Telangana State Agricultural University, Rajendranagar, Hyderabad, India*

[3]*Krishi Vigyan Kendra, Kalyandurg, Anantapur, Andhra Pradesh, India*

Abstract

Agrarian innovation is profoundly gendered, from its beginning in investigation and improvement to its dissemination, get to, and appropriation. Innovation can be pivotal in supporting women's interest in higher-value exercises in worldwide agri nourishment chains and in cultivating environmentally maintainable rural homes fundamental to managing the impacts of climate alter and moving forward the employments of provincial women who are among those most helpless to climate impacts. Be that as it may, women confront numerous gender-based boundaries to getting to advances, and accessible advances regularly fall flat to meet their specific needs. To address these sexual orientation lacunas in agrarian innovation development and get to think about prescribes a gender-sensitive approach to be received from the beginning stages of inquire about and improvement, preparing and expansion administrations, as well as credit arrangement and measures supporting women's possession and control of innovations. Sexual orientation contemplations ought to advise arrangements and investigate needs in arrange to guarantee that the diverse financial parts and challenges of both men and women are suitably recognized. Toward these closes, it is vital to bolster capacity-building in sex-disaggregated information collection within the agrarian segment. These measures would be vital steps toward tending to the 2030 Economical Advancement Objectives, counting nourishment security, maintainable agribusiness, sexual orientation uniformity, and strengthening of all young women.

Keywords: *Women empowerment, Technology, Climate impacts, Gender.*

11.1 Introduction

Gendered Advancements utilize strategies of sex and intersectional investigation to form modern information. It looks to saddle the inventive control of this sort of examination for advancement and revelation. The agent's address is, do considering these components include important measurements to investigate? Do they take research in unused bearings? To reply to these questions, Gendered Developments creates down-to-earth strategies of sex, sexual orientation, and intersectional examination for researchers and gives case ponders as concrete outlines of how this sort of investigation leads to revelation and advancement.

Agrarian developments tend to affect women and men in an unexpected way. Most developments center on settling specialized issues, such as raising yields, withstanding natural stretch, and overseeing destitute soil or seeds. Such advancements frequently come up short of accounting for 1. How sex standards impact the usage of specialized arrangements or 2. How the execution of specialized arrangements will impact sex standards. Thus, developments may not reach women and may indeed intensify sex imbalance (Rola-Rubzen *et al.*, 2020; Quisumbing *et al.*, 2014; Uncivilized et al., 2017; Bock & van der Burg, 2017).

Investigate appears that sexual orientation and other standardizing social structures frequently lead to unequal benefits from advancement openings. Prohibitive sexual orientation standards are profoundly engrained in teaching, such as the family, instruction, keeping money, and so on. These oblige the organization of women, their get to assets and administrations, portability, representation, and decision-making capacity, particularly for less advantaged bunches (Cole *et al.*, 2015; Badstue *et al.*, 2018). Typically the case around the world (Bock & Shortall, 2017).

Sexual orientation specialists over the Consultative Group on International Agricultural Research (CGIAR) made Empowering sex correspondence in agrarian and natural advancement (Gennovate), pointed at creating sex transformative approaches that insert sexual orientation norms in advancement forms. These incorporate: 1. cultivating basic examination of sexual orientation parts, standards, and relations; 2. recognizing and fortifying positive standards that back balance; 3. advancing the position of women, young women, and marginalized bunches; 4. changing fundamental social structures; arrangements and broadly held convictions that sustain sex imbalance. Petesch *et al.*, 2017, 2018, 2018a, 2018b; Badstue *et al.*, 2018; Danielsen *et al.*, 2018, with reference to IGWG 1997; see too Hillenbrand *et al.*, 2015; Aregu *et al.*, 2019; van der Burg, 2019; Fischer *et al.*, 2019.

11.2 Gendered Developments

11.2.1 Gill nets in Bangladesh

To overcome nourishment destitution, World Fish led the USAID-funded venture Aquaculture for Wage and Sustenance with nearby accomplices in Bangladesh. World Fish is one of the institutes of the around the world Consultative Group of International Agricultural Research Institutes (CGIAR). The project combined the presentation of unused innovation with a gender-transformative approach (Kruijssen *et al.*, 2016; see moreover WorldFish, "Gill nets enable women to gather nutritious fish in Bangladesh"'

Fish is the foremost imperative animal-source nourishment in Bangladesh. Roughly 60% of the populace eats fish at least each other day, with every day per capita utilization at 44 grams for the poorest family units. In 2012, around 4.27 million family units (20% of rustic family units) had fish lakes (Belton & Azad, 2012).

Gender-related social and devout desires disallow women from collecting fish indeed from their claim lakes. Such assignments are seen as the obligation of men. Women are hesitant to enter lakes since they get their sarees wet, which ought to be washed at that point. As a result, women generally bolster the fish and offer assistance clean the lakes, whereas men gather and showcase the fish, and make budgetary choices for the family. To back sex equality and to guarantee nourishment security, World Fish begun to present mola gill nets within the poorest parts of Bangladesh. Gill nets are lightweight and sufficient for women to cast from the bank of the pond without getting into the water. Gill nets come in networks of distinctive sizes with work estimate deciding the sort of fish caught. A fish swims into the net but passes as it were portion way through the mesh. When it battles to free itself, the netting slips behind the gill cover to trap the fish.

Since the nets are utilized to capture as it were nutrient-rich mola fish, there's a small threat of netting larger fish, which men tend to control. Women went to an eight-day preparation course on how to form and utilize a gill net. The nets were 7.5 meters wide by 1.8 meters tall. Women were instructed to create their claim nets by buying low-cost mesh and rope from the showcase and utilizing purge bottles (for drifts) and rocks or broken bricks (as sinkers). Collecting fish with these nets takes around 30 minutes from setting the net to pulling within the capture.

11.2.2 Nano Ganesh -Remote control for water pump in India

Tending to imperatives on women's portability, Nanao Ganesh was presented. Portable phones can be utilized to control apparatus remotely. Nano Ganesh may be a remote control for water pumps created by an Indian company, Ossian

Agro Robotization. Its electronic equipment for turning pumps off and on can be enacted remotely by versatile phone. This makes a difference for women (and men) agriculturists to utilize water more efficiently. Without remote controls, agriculturists either need to make extraordinary trips to the fields at night to turn pumps on (power is often available as it were amid off-peak hours), or they have to be take off the pumps on to run on the irregular power supply, squandering water, diminishing pay, and disintegrating soil. Women agriculturists, in particular, find the nighttime trips unsafe and troublesome. This development has moreover created modern exercises, which give extra sources of income for women within the company's rustic call centers, gadgets gathering, and in promoting and preparing.

11.2.3 Mechanized potato selection and gender roles in Bolivia

Analysts from the Latin American sexual orientation group of the Worldwide Potato Center (CIP) in Lima, Peru, needed to get what underlies innovation appropriation choices for men and women in high-elevation Andean communities where cultivating frameworks are based on potato development. The analysts from CIP, which is portion of the Consortium of Worldwide Rural Centers, found that innovation selection choices are the result of the complex interaction of numerous components, counting the characteristics of the innovation itself and sexual orientation parts. In turn, the advances that are eventually received may carry forward sex predispositions on the off chance that wants of users are not taken into consideration at all stages.

For illustration, in Jacopampa, the Plurinational State of Bolivia, women are dependable for physically sorting and selecting potatoes by measure. But in spite of the fact that hardware to spare time by mechanizing this errand was presented, it was never really embraced. The women who were the focused on users of the innovation, in truth, were not expressly included within the collection of input and exhibit of the modern innovation. More men than women taken an interest in approval gatherings approximately the modern gear, and they were able to function the hardware without trouble. In any case, the gear was not suited to the physical prerequisites of women. As a result, women found the hardware too tall to function, and had trouble lifting the overwhelming packs of potatoes to bolster into the machine. Once this obstruction to appropriation was caught on, a less difficult, gravity-fed, low bearing machine was effectively presented.

11.2.4 Women's empowerment in agriculture and innovation in Southeast Asia

How can women's strengthening lead to way better appropriation of agrarian and cultivating innovation? That address propelled a consideration by Akter *et al.* (2017) of 12,000 cultivating families in Southeast Asia that assessed women's empowerment by employing a measure called the Women's Strengthening in Horticulture List. The list considers a arrangement of variables to degree strengthening, counting information on get to to expansion administrations (training, agricultural information, and agricultural technologies).

The division of cultivating errands between men and women within the locale illustrates common designs. Men rule decision-making over most assets, whereas women's interest in key choices changes over the locale. For the case, the study found that the biggest degree of women's inclusion in decision-making is within the Philippines. Most rice cultivating choices are made together by spouses and spouses, and women play vital parts in community and rural bunches. Women, moreover, take part more effectively than men in most agrarian gatherings organized by the nearby expansion office, and their spouses tune in to the data they pass on.

In Myanmar and Indonesia, on the other hand, women are underrepresented in formal groups, which leads to a need for expanded administrations. Women in these two nations are sharp at taking part in gatherings and preparation, but solicitations are ordinarily extended to men.

Concurring to the ponder, engaged women have get to more systems and data, counting around available technologies, that enable them to make more educated production decisions for their ranches and advocate for measures that would progress family livelihoods and welfare. This finding shows the significance of tending to sexual orientation holes in agribusiness through country-specific sexual orientation mediations.

11.2.5 Gender impacts of participatory plant breeding programs in West Africa and Rawanda

Participatory plant breeding programs involve farmers straight forwardly in plant breeding and making unused plant assortments. Analysts can draw on farmers' information and involvement of neighborhood conditions, whereas agriculturists can select and adapt edit assortments to their particular situations, agrarian hones, and needs. Numerous studies have discovered that participatory plant breeding leads to speedier selection and higher yields since it unequivocally joins and reacts to the interests and needs of different clients. In any case, numerous considerations moreover show that the benefits for women are constrained unless these programs consolidate an unequivocal sex measurement and address women's particular imperatives, such as the need for portability and transport, time burden, requirements for information and preparation, and sexual orientation standards.

Women are regularly central to neighborhood rural development frameworks, especially with regard to plant breeding strategies, era of unused plant assortments, and serving as stores of information around local varieties and genetic data. In different parts of the world, women are capable for sorting and selecting seeds for distinctive crops. In a study of over 150 participatory plant breeding ventures, Ashby and Lilja (2004) found that counseling women and including them in varietal assessment driven to superior ac ceptability and speedier appropriation. Ceccarelli and Grando (2007) conclude that participatory plant breeding gives voice to agriculturists and marginalized women in combining logical revelation and conventional information. Participatory plant breeding approaches that include women within the advancement of NERICA (or New Rice for Africa, an unused high-yield rice assortment) in West Africa and beans in Rwanda have driven way better-performing assortments and upgraded benefits for women.

(*Sources:* ACET, ed. (2017); Meinzen-Dick *et al.* (2010); and Ragasa *et al.* (2014).

11.2.6 Communal leasehold ranger service courses of action and labour-saving ranger service and animals hones in Nepal

In 2004, Nepal embraced an inventive organization course of action that permitted the expansion of tree crops and animals without debasing the woodland. The program has too given vital changes to women's employments. Through the Leasehold Ranger Service and Animals Program, communities within the Himalayan foothills were allowed 40-year renewable leases over the timberland, and unused crops and goats were presented alongside economical generation hones and commerce and showcasing methods. The program given women - who are regularly landless and spend much time each day collecting water, creature grub, or wood for fuel - with get to to arrive where they might raise their goats and plant reasonable grasses, vegetables, and grub trees.

The Nourishment and Agribusiness Organization of the Joint Together Countries has given specialized help and capacity-building to a number of ventures that advanced communal ranger service lease holding. In specific, women received a sort of grass that was utilized for making brooms, the deal of which given an extra source of wage. The grass was used for scrounging animals, and the dried stems as sticks to bolster developing vegetables.

Women utilizing the relegated arrive (instead of strolling long separations looking for ground grasses) detailed critical time investment funds (between 3 and 10 hours per day). This empowered them to lock in in other income-earning exercises. The case of Nepal shows how a gender-sensitive procedure centering on getting arrival and animal resources, in expansion to preparing, can upgrade women's appropriation of technologies. Sources: FAO (2019a); and Shapiro *et al.* (2015).

11.2.7 "Banish the hand hoe to the museum:" Sustainable agricultural mechanization in Africa:

An activity to advance feasible rural mechanization in Africa beneath the motto to "banish the hand digger to the gallery by 2025" challenges the cliche picture of an African lady developing her to cultivate with a basic hand tool. This picture reflects the reality of moo levels of mechanization in Africa. There are 5 tractors for every 1,000 ranchers in Africa, compared with nearly 1,600 tractors for every 1,000 ranchers within the United States. This can be compounded by gender-based norms that consign women to the labor-intensive errands of weeding alongside gender-based imperatives on their get to mechanized hardware. For illustration, in a few parts of Africa, it is considered unseemly for women to work with bulls or to drive tractors (Ragasa, 2012).

Economical mechanization may be a implies to diminishing the burden of difficult work, soothe work deficiencies, progress the efficiency of rural operations, and improve benefits and advertising get to. Maintainability alludes to the technology's financial, social, natural, and social measurements. A public-private partnership between the Collusion for a Green Transformation in Africa and the Renting Leveraging Jobs Consortium case seeks to overcome the two key obstructions to mechanization in African farming - the need for financing and the need to prepare - by coupling lease-to-own financing with preparing the on how to utilize the hardware. Sources: USAID (2016); and Nourishment and Agribusiness Organization, (http://www.fao.org/sustainable-agricultural- mechanization/overview/what-is-sustainable-mechanization/en/)

Table 11.1: Summary of Gender-Sensitive Agricultural Practices and Outcomes

Region / Country	Innovation / Study	Gendered Impact / Outcome	Source Citation
Bolivia	Mechanized potato selection	Gender bias in adoption due to physical requirements not met for women	N/A (Example)
Southeast Asia	Women's Empowerment Study	Women's empowerment linked to better technology adoption and decision-making	Akter *et al.* (2017)
West Africa, Rwanda	Participatory Plant Breeding	Enhanced variety acceptance and adoption through women's involvement	Ashby & Lilja (2004), Ceccarelli & Grando (2007)
Nepal	Communal Leasehold Forestry	Increased access to resources and income through gendersensitive approaches	FAO (2019a); Shapiro *et al.* (2015)
Africa	Sustainable Agricultural Mechanization	Challenge to gender norms; increase in mechanization to alleviate women's hard labor	Ragasa (2012); USAID (2016)

11.3 Techniques and approaches to improve women's get to and utilize of agrarian innovations counting on and off-farm time and labour-saving advances

i. Great hones to move forward women's get to to enlisted work, agrarian inputs and to move forward cooperation in farmers' organizations
ii. The farming esteem chain considers getting it what works to support women farming business people in growing their businesses and to support youthful women in locking in in farming esteem chains as wage workers and business people.
iii. Imaginative approaches to make strides women's get-to-budgetary administrations within the farming segment
iv. Approaches to move social standards influencing women's interest and get to to rural services and their capacity to form choices on agribusiness generation and salary
v. Arrangement of sex touchy rural admonitory administrations and procedures that offer assistance to bridge the sexual orientation computerized separate
vi. Considers on affect of existing corporate hones to move forward farming business conditions for women agriculturists
vii. Procedures to address unpaid care and residential work influencing women's interest in horticulture.
viii. Procedures to reinforce women's arrive rights.
ix. Affect of rural programs on women's strengthening that can offer assistance to recognize key regions of disempowerment

Future policies and venture programs ought to center on unpaid care and residential work (UCDW), which is the often-invisible work of caring for children, elderly and debilitated individuals, and those living with incapacity, as well as cleaning, cooking, washing, collecting water, and bringing kindling, among other assignments. Conventional work is characterized as work that's profitable and conveys a reasonable pay or least wage, ensures security within the work environment and social security for families, superior prospects for individual development and social integration, flexibility for people to precise their concerns, organize and take part in the choices that influence their lives and uniformity of opportunity and treatment for all women and men.

11.4 Conclusion

The integration of gender and intersectional analysis into agricultural innovation is not just beneficial but essential for creating sustainable and equitable agricultural practices. Research such as Gennovate and the work of gender specialists within CGIAR underscore the importance of transforming how agricultural innovations are developed and implemented. By actively challenging and reshaping gender norms and structures that limit women's access to and control over resources, these approaches aim to ensure that both men and women can equally benefit from agricultural advancements. This transformative strategy not only promotes gender equality but also enhances the overall effectiveness and sustainability of agricultural interventions, leading to more resilient rural communities and contributing to broader developmental goals.

References

Akter S., Rutsaert P., Luis J., Htwe N.M., San S.S., Raharjo B. and Pustika A. (2017). Women's empowerment and gender equity in agriculture: A different perspective from Southeast Asia. Food Policy. 69:270-279. Available at https://doi.Org/10.1016/j.foodpol.2017.05.003.

Aregu, L., A. Choudhury, S. Rajaratnam, M. van der Burg and C. McDougall (2020). Implications of agricultural innovation on gender norms: Gender approaches in aquatic agriculture in Bangladesh. In Sachs, C. (Ed.). Gender, Agriculture and Agrarian Transformations: Changing Relations in Africa, Latin America and Asia. Milton Park, Abingdon, Oxfordshire: Routledge, 162-179.

Ashby J. and Lilja N. (2004). Participatory Research: Does It Work? Evidence from Participatory Plant Breeding. Paper presented at the 4th International Crop Congress: New directions for a diverse planet. Brisbane, Queensland, Australia, 26 September to 1 October.

Badstue, L., Petesch, P., Feldman, S., Prain, G., Elias, M., & Kantor, P. (2018). Qualitative, Comparative and Collaborative Research at Large Scale: An Introduction to GENNOVATE. Journal of Gender, Agriculture and Food Security, 3(1): 1-27.

Belton, B., and Azad, A. (2012). The characteristics and status of pond aquaculture in Bangladesh, Aquaculture, 358: 196-204.

Bock, B., & van der Burg, M. (2017). Gender and international development. In: Bock, B. & Shortall, S. (Eds.) Gender and rural globalization. Wallingford and Cambridge, MA: CABI, 245-252.

Bock, B., & van der Burg, M. (2017). Gender and rural globalization. Wallingford and Cambridge, MA: CABI.

Burg, M. van der. (2020). 'Change in the Making': 1970s and 1980s building stones to Gender Integration in CGIAR Agricultural Research. In Sachs, C. (Ed.). Gender, Agriculture and Agrarian Transformations: Changing Relations in Africa, Latin America and Asia. Milton Park, Abingdon, Oxfordshire: Routledge, 35-57.

Ceccarelli S. and Grando S. (2007). Decentralized-participatory plant breeding: An example of demand driven research. Euphytica. 155: 349-360

Cole, S.M., Puskur, R.,Rajaratnam, S. & F. Zulu, F. (2015). Exploring the intricate relationship between poverty, gender inequality, and rural masculinity: A case study from an aquatic agricultural system in Zambia. Culture, Society and Masculinities, 7(2): 154-170.

Danielsen, K., F. Wong, and M. Mukhopadhyay (2019). Integrating "gender" in research for development: How you interpret the term can shape project achievements. CIMMYT gender resources for agricultural research and development professionals. CDMX, Mexico: CIMMYT.

FAO (2019a). Fostering the Uptake of Labour-saving Technologies. (Food and Agriculture Organization of the United Nations Rome). Available at http://www.fao.org/3/CA2731EN/ca2731en.pdf

Fischer, G., Wittich, S. & Frundt, S. (2019). Gender analysis in farming systems and action research: A training manual. Ibadan, Nigeria: IITA.

Hillenbrand, E., Karim, N., Mohanraj, P. & Wu, D. (2015). Measuring gender transformative change: A review of literature and promising practices. CARE USA. Working Paper.

Kruijssen, F., Audet-Belanger, G., Choudhury, A., Crissman, C., Dalsgaard, J. P. T., Dawson, C., Dickson, M., Genschick, S., Islam, M.M., Kaminski, A., Keus, H.J., McDougall, C. Banda, L.E., Muyaule, C. & Rajaratnam, S. (2016). Value chain transformation: Taking stock of WorldFish research on value chains and markets. Penang, Malaysia: CGIAR Research Program on Aquatic Agricultural Systems. Working Paper: AAS-2016-03.

Lawless, S., Cohen, P., McDougall, C., Orirana, G., Siota, F., & Doyle, K. (2019). Gender norms and relations: implications for agency in coastal livelihoods. Maritime Studies 18, 347-358.

Lawless, S., Doyle, K., Cohen, P., Eriksson, H., Schwarz, A. M., Teioli, H., Vavekaramui, A., Wickham, E., Masu, R., Panda, R & McDougall, C. (2017). Considering gender: Practical guidance for rural development initiatives in Solomon Islands. Penang, Malaysia: WorldFish. Program Brief.

Meinzen-Dick, R., Quisumbing, A., Behrman, J., Biermayr-Jenzano, P., Wilde, V., Noordeloos, M., Ragasa. C. & Beintema, N. (2010). Engendering agricultural research. International Food Policy Research Institute (IFPRI) Discussion paper 973.

Petesch, P., Badstue, L. & Prain, G. (2018). Gender norms, agency, and innovation in agriculture and natural resource management. The GENNOVATE methodology. Mexico, D. F.: CIMMYT.

Petesch, P., Badstue, L., Prain, G., Elias, M., & Tegbaru, A. (2017). Entry points for enabling gender equality in agricultural and environmental innovation. GENNOVATE resources for scientists and research teams. CDMX, Mexico: CIMMYT.

Petesch, P., Badstue, L. B., Camfield, L., Feldman, S., Prain, G., & Kantor, P. (2018a). Qualitative, comparative and collaborative research at large scale: the GENNOVATE field Methodology. Journal of Gender, Agriculture and Food Security, 3(1): 28-53.

Petesch, P., Bullock, R., Feldman, S., Badstue, L., Rietveld, A., Bauchspies, W., Kamanzi, A. Tegbaru, A. & Yila, J. (2018b). Normative dimensions of agency and agricultural decision-making in Sub-Saharan Africa. Journal of Gender, Agriculture and Food Security, 3(1): 108-130.

Quisumbing, A., Meinzen-Dick, R., Raney, T.L., Croppenstedt, A. Behrman, J. A. & Peterman, A. (Eds.) (2014). Gender in agriculture. Closing the knowledge gap. Dordrecht: Springer.

Ragasa C, Sengupta D, Osorio M, Ourabah Haddad N and Mathieson K (2014). Gender-specific Approaches and Rural Institutions for Improving Access to and Adoption of Technological Innovations. Food and Agriculture Organization of the United Nations. Rome.

Rietveld, A., & Farnworth, C.R. (2018). Towards gender-responsive banana research for development in the East-African Highlands. GENNOVATE resources for scientists and research teams. CDMX, Mexico: CIMMYT.

Rola- Rubzen, M. F., Paris, T., Hawkins, J., & Sapkota, B. (2020). Improving Gender Participation in Agricultural Technology Adoption in Asia: From Rhetoric to Practical Action. Applied Economic Perspectives and Policy, 42(1): 113-125.

Shapiro B., Adhikari B.R., Baral N.R., Hancock B., Kafley G., Koirala P. and Reijmerinck J. (2015). Regenerating Forests and Livelihoods in Nepal: A New Lease on Life - Unfolding the Experience of 20 Years of Poverty Alleviation through Leasehold Forestry in the Himalayas. CABI. Wallingford, UK.

USAID (2016). New model aims to banish the hoe to the museum. U.S. Agency for International Development. Washington DC. 11 November. Available at http://agra.org/sstp/news/new-model-aims-to-banish-the-hoe-to-the-museum

12

Empowering Changemakers Cultivating Women's Leadership in Technology

Vineeta Chandra[1] and Sameer Mahapatro[2]

[1]*Department of Agricultural Extension Education, Centurion University of Technology and Management, Paralakhemundi, Odisha, India*

[2]*Department of Agronomy, Centurion University of Technology and Management Paralakhemundi, Odisha, India*

Abstract

Female entrepreneurship is one of the main forces behind social innovation in the nation's developing economies. However, these entrepreneurs need access to development ecosystems for growth and scale to effect systemic change. Thus, this contribution aims to provide instances of current support systems that enable the nation's female social entrepreneurs to expand their enterprises. In nearly every aspect, women are falling short of men. This disparity widens when one considers the digital sphere and digital careers, where women are underrepresented in accessibility, affordability, and ICT-based education relative to men. In light of this, the chapter makes an effort to provide enough information on this requirement for women's empowerment by using various national and international data and asking questions about many aspects of the digital gender divide. The final section of the chapter discusses the next steps that will bring about a society that is both sustainably developed and more gender inclusive in the digital sphere.

Keywords: *Female entrepreneurship, women's empowerment, ICT-based education, Digital career.*

12.1 Introduction

Technological advancements have completely changed human existence, bringing about profound societal shifts and stimulating the economy. Empowering women will have a good impact on women's daily lives and is essential to the growth of any society. Women now have many possibilities to showcase their creativity, ideas, and technological prowess both in India and around the world thanks to the advancements and inventions in the field of information technology. This review's main goal is to investigate how women use digital technology in urban and rural areas. India is rapidly digitizing, embracing all facets of technological advancement, and as a result, offering plenty of job prospects for women.To reduce the disparity between genders in technology fields in Indian society, women's involvement in these fields, both in rural and urban areas, must be the main focus. Education helps rural women not only become more digitally literate but also more economically independent. Digital literacy's main focus, towards digitization, is the most crucial ability needed for women's empowerment. The NSSO survey indicates that compared to urban women, rural women have a substantially lower literacy rate. Because tech education is not widely available, there is a greater disparity in the use of technology between women in urban and rural areas. Following globalization, technology has always been a field dominated by men. However, this trend is gradually moving towards including urban women; however, it should also reach out to rural women, as it may empower them in all areas. The main determinants of how women can benefit from technological innovation and reach higher levels are their literacy and social standing. Women in rural and urban areas have been updated by the growth of mobile technology, which has been a beneficial movement to make women aware of global trends. However, there are now obstacles to inclusive growth due to the quick development of electronic technology. Women are underrepresented, for example, in economically important industries such as technology and communications technologies (Pappas *et al.*, 2017).

The underrepresentation of women in the workforce is a real problem in the vast terrain of today's computer business. In addition to being a matter of equity or equality, the gender gap obstructs opportunities for creativity. A rural lady named "Pooja" wanted to demonstrate her designing skills, and the internet helped her achieve this, according to Notermans (2019). Giving these brilliant women the chance to succeed would help them become role models for additional rural women who wish to improve their own digital literacy. Updating the industry's credentials is strategically important to attain comprehensive expansion. One of the main issues in minimizing the gender gap and bringing up important issues that impact women's life is the extremely low participation of women in the IT industry and start-ups. These worries call for a reassessment of the IT sector with an eye toward diversity and women's enrolment, which can result in a safe and diverse workplace. According to Nasscom (2021) figures, approximately 36% of women labor in the tech sector, but just 7% of them hold lead positions in the corporate hierarchy. Skillsoft's (2022) estimates show that just 13% of women hold director-level jobs, while approximately 17% have management positions in the computer industry. The employment of women as scientists, engineers, and technologists poses a significant challenge to the gender gap, as only 14% of women who graduate from STEM (science, technology, engineering, and mathematics) programs do so out of 43% of all women who graduate from STEM programmes (World Bank, 2022). A Harvard Business Review research states that less than 3% of venture capital investments go to start-ups that women exclusively run and that less than 15% of cheque writers are female. Concerns about bias and discrimination have also been raised by the use of AI and machine learning, with multinational corporations replacing female employees with male ones. The gender gap in technological and business domains has been caused by the structural construction of developing technologies, resulting in limited funding for women-led start-ups. The primary goal of preventing the gender gap in technology can be accomplished by enabling rural women to rely on technology independently and educating them about various "Digital India" programmes that advance women's economic and social standing in India. Using information and communication technologies, digitization will open up new possibilities for women to increase their entrepreneurship chances (Rajahonka and Kaija, 2019). Women's empowerment can assist them in enhancing their participation in digitization processes, which will help them make better decisions and overcome hurdles in the modern day's social, political, and economic spheres (Cummings and Tam O'Neil, 2015). Since IT offers flexible workat-home options, women can confidently pursue careers in this field (Malhotra, 2015).

12.2 Women and Technology

Technology-related careers require competencies not typically associated with women (Mora, 2015). Women make up 50% of university students in most developed countries, yet they are underrepresented in STEM (science, technology, engineering, and mathematics) fields (Waaijer *et al.*, 2016). For younger children, there is no gender difference in computer use, but by the time they reach 16 or 17, the disparity becomes more pronounced (Dixon *et al.*, 2014). It may be because men write computer software that boys generally find computers more attractive and become more comfortable with them. Moreover, women think they are less talented than males, even when there is little to no difference in their skill sets. (Dixon *et al.*, 2014; van Dijk, 2017).

Better arithmetic skills in girls make them less likely than less proficient boys to pursue STEM fields in college, which lends credence to the idea that there are "leaks" in the pipeline leading to education for women (Ferguson, 2016). Research indicates that some males respond to societal beliefs that computer science is a topic better left to men by "chasing" girls and women away from the field (Michell *et al.*, 2017). As a result, according to (Pappas *et al.* 2017), In Europe, women account for only 30% of ICT workers and have created 9% of ICT apps. This runs counter to studies that demonstrate women are more creative than men and that having more women in leadership roles leads to better decision-making (Pappas *et al*, 2017).

12.2.1 Women as managers and entrepreneurs in a digital context

According to Joshi *et al.* (2015), women are underrepresented in leadership positions within organizations and are paid less and promoted less frequently than men. In the European Union in 2016 there were only thirty-three percent female managers (European Union, 2017). Managers had the biggest disparities in hourly wages among professions, with women earning 23% less than males (European Union, 2017). Women deal with "sticky floor" and "glass ceiling" concerns simultaneously; while less educated women struggle with sticky floors, better-educated women face challenges as a result (Ahmad & Naseer 2015). It has been proposed that being more qualified may be a woman's best chance, even though education may not always help women in the same ways as men (Ahmad & Naseer, 2015; Mora, 2015).

Furthermore, bias may be encountered by women pursuing jobs in domains where men predominate because they do not "fit" the stereotypes and are thought to be less competent or successful than men (Joshi *et al.*, 2015). Poor leadership qualities are one example of this. Furthermore, female peers' or superiors' efforts may go unappreciated. Joshi *et al.* (2015) found that across a wide range of industries and occupations, gender inequalities in performance evaluations were almost 14 times smaller than gender differences in organizational rewards. There have also been claims that due to the general perception that women are less risk-taking than men, those who take chances may encounter prejudice or be labeled as "cocky" (Jewell, 2011). Considering all the, women's unusually high organizational positions have huge symbolic significance for other women (Jewell, 2011; Joshi *et al.*, 2015).

Men and women in the EU make roughly 16% less money on average, but women's income is highly influenced by computer skills, which are slightly higher (5.3% for men and 6% for women) (European Union, 2017; Pappas *et al.*, 2017). Therefore, learning digital skills could be seen as especially beneficial for women in their careers.

About 30% of European entrepreneurs are women, according to the Statistical Data on Women Entrepreneurs in Europe (European Commission, 2014; Pappas *et al.*, 2017). It has been said that to break through the "glass ceiling," women are starting their own enterprises at an increasing rate. Women launch almost 60% of new enterprises in Canada and Norway, demonstrating this effect (Afrah & Fabiha, 2017). Furthermore, in Europe, women make up most one-person firms (78%) (Pappas *et al.*, 2017). Digital innovations are changing the face of business and creating new opportunities. CT use and digitalization, therefore, provide women entrepreneurs with equal access to firm growth prospects with men as well as increased opportunities to grow both themselves and their businesses (Afrah & Fabiha, 2017; Pappas *et al.*, 2017; Shah & Saurabh, 2015). Scuotto *et al.* (2019) define a "Female Digital Entrepreneur" as a woman who "explores market opportunity by exploiting the digital space to create something new." This idea has even inspired new concepts. Opportunities for global value chain entry for entrepreneurs, particularly women entrepreneurs.

Nonetheless, women encounter additional difficulties in their professional lives. It has been observed that female entrepreneurs face more obstacles in obtaining finance, such as start-up capital, than their male counterparts. These obstacles typically stem from insufficient time or personnel. Furthermore, women are more likely than males to value their families. Furthermore, women's capacity to progress in their occupations is restricted because they are still expected to perform most of the jobs related to social reproduction, such as cooking, cleaning, caring for young children, and caring for aged family members (Jewell, 2011). Data from 2016 illustrates this impact within the European Union: 79% of women cooked or cleaned the house, and 92% of them took care of their children on a daily basis, compared to 34% and 68% of men (European Union, 2017).

According to data from the European Union (2017), women made up 32% of the working population in the region in 2016 compared to 9% of men, while 84% of men and 55% of women with three or more children were employed. Furthermore, it has been claimed that women are drawn to a variety of "non-standard working arrangements," such as hybrid entrepreneurship, which blends employment and entrepreneurship (Solesvik, 2017).

Digital technology have generally increased flexibility in the workplace and made it more difficult to define the boundaries between work and play (Gronlund & Oun, 2018). This presents opportunities and difficulties for women in balancing their duties to their families and careers. The work-life balance of women may be improved by the growing usage of digital technologies in combination with flexible work schedules. It has been demonstrated, for example, that women who successfully organize and manage their family problems using their mobile phones do so while working (Roy, 2016). But in our contemporary, flexible work environment, people are expected to be able to define and organize their own work and draw boundaries between their employment and personal life. Given that studies show that women really have less influence over their work and schedules than men do, one may claim that modern working life has not yet caught up to the expectations of modern women (Gronlund & Oun, 2018).

Technologies are tools or means to do certain tasks that need to be integrated into social contexts, many of which are dynamic and gender-related. As such, they have no inherent value. According to (Huyer and Sikoska 2003), women need to be convinced of the advantages of ICTs. Because they are not gender-neutral, ICTs nowadays are also not always appropriate for women's needs (Huyer & Sikoska, 2003), at least not without adequate adaptation.

12.2.2 Technology Diffusion and Domestication

Technology is a major driving element behind innovation. Innovation research has looked at various subjects, including the diffusion of innovation, or the process by which innovations are adopted across a network of businesses (Rogers, 1995). Usually, a curve illustrates this process, with different user types embracing innovations at different phases (Meade & Islam, 2006). Adopting new technology, however, would not benefit the adopting companies because it needs to be linked with existing activities (Hazen & Byrd, 2012; West & Bogers, 2014). It takes both technological know-how and a cooperative culture to adapt innovations to the organization's unique circumstances (West & Bogers, 2014).

Opponents of diffusion theory argue that the approach ignores the ways in which individuals modify or change technologies (Deidre, 2007). The social construction of technology theory holds that rather than being "trickled down" or dispersed, technology is socially produced to fit different social circumstances and transforming society and its surroundings (Harwood, 2011). For example, gender affects the development and understanding of technology, but technology also affects gender roles. For example, computer use was traditionally associated with men, but this changed as more women started using them. Consequently, the gender gap in overall computer and Internet use that had been noted in the late 1990s had diminished by the late 2000s (Dixon *et al.*, 2014).

Technology theory Domestication, which is based on the idea of contexts, emphasizes how users contribute to innovation and their efforts to incorporate technology into daily life. Domestication theory describes how inventions—especially new technology—are "tamed" or modified by users to better suit their needs. The domestication process consists of four stages: conversion, appropriation, objectification, and assimilation. Ownership is crucial during the appropriation stage, which occurs when a technology is offered to a user. During the objectification stage, the user makes temporal and spatial adjustments to grasp the new technology's value. When the thing is incorporated, its use is emphasised. The conversion phase focuses on the user's impressions of the technology and how they describe their interactions with it (Deidre, 2007; Silverstone *et al.*, 1992).

A deeper understanding of the creation, uptake, application, and evolution of technologies is necessary to apply constructive and domesticated strategies. Additionally, they question the traditional division of labour between active producers and passive users (Harwood, 2011). According to these approaches, using technology involves both social interaction and knowing how to use it. The "struggle" to make the technology work involves some inventiveness (Harwood, 2011).

12.3 Ways to Avoid Gender Gap in Technology: Rural women in digital era

Women now have the opportunity to completely change their lives through digitalization by utilizing technology more effectively from home and shattering all restrictions, according to reports from Women (2019). Men make up the majority of internet users (43%), with women owning mobile devices and using the internet at just 28%. When this disparity was evaluated at 114 million, it was found that women were never taught how to read or write numbers or distinguish any of the keys on mobile phones. They had never been trained to decipher texts, so they depended on their husbands to use the equipment. Their cell phone was only ever used for taking calls. The comparatively minor gender disparity in mobile phone usage can be explained by the higher percentage of literate women and increased exposure to advertising in metropolitan areas as compared to rural areas. These factors also lessen societal barriers. (Keeley, 2017).

12.4 List of Technology Schemes to Empower Women In India

12.4.1 The SUNIL programme, which aims to strengthen, upscale, and nurture local innovations for livelihood

With the help of technological interventions that meet their social requirements, this program aims to improve women's livelihood standards by providing them with enterprise efficiency and technology training. The capacity building of community-based organizations, non-governmental organizations, knowledge institutions, and social start-ups in India is facilitated by this technology training.

12.4.2 Science & Technology (S&T) for Women

Scheduled Tribe (ST) and Scheduled Caste (SC) populations are being supported by the creation of centers for science, technology, and innovation (STI) around the country. Raising living standards and creating sustainable livelihoods, these hubs support developing, enhancing, and implementing pertinent and appropriate STI techniques to support the improved growth of the SC and ST populations. Twenty STI Hubs (seven for ST and thirteen for SC) have been constructed across the country's different regions in order to improve livelihood systems and boost their resilience through S&T interventions.

Over 50,000 women nationwide have benefited almost entirely from the 1982-launched "Science and Technology for Women" program, which aims to empower women via science and technology and advance gender equality. Approximately 1500 projects across various industries targeted at enhancing women's lives locally had received funding. Over 20,000 women benefited from the establishment of about 46 women's technology parks and acted as one-stop shops where various technologies could come together and form integrated forward and backward links. These developments helped women acquire new skills and knowledge in science, which helped them become successful entrepreneurs. This initiative provides primary women beneficiaries for projects pertaining to creating women's technology parks, the decrease of drudgery, health and nutrition, the elimination of occupational dangers, and enhancing livelihood status.

12.4.3 Women Technology Park (WTP)

Women Technology Parks (WTP) combine science, technology, and innovation to improve the weakest link in a community's prevalent women's livelihood system. They foster social entrepreneurship and women's employment depending on the strongest relationship. To ensure local self-reliance for economic regrowth, efforts are made to guarantee that women in rural and peri-urban areas receive training in WTPs for skill development and capacity building in accordance with the National Skill Qualification Framework through S&T-based interventions, using resources specific to the area. There are currently 25 Women Technology Parks operating around the nation. These WTPs trained close to 1500 women in the preceding year through offline and online training sessions.

12.4.4 TAP-RISE, the Technology Acceleration Platform for Social Entrepreneurship and Rural Innovation

i. The SUNIL program, aims to strengthen, upscale, and nurture local innovations.

ii. For livelihood. TIDE, or Technology Interventions for the Elderly and Disabled Young Scientists and Technologists Scheme (SYST).

iii. Tribal Sub Plan (TSP) and Scheduled Caste Sub Plan (SCSP).

iv. Women in Science and Technology (S&T for Women).

v. Technology Parks for Women.

vi. Resource Centre for Community Resilience (CRRC).

12.4.5 MeitY Digital India Internship Scheme 2023

12.4.5.1 Purpose

Through an internship, a student can gain real-world experience in the workplace while being supervised by a skilled and knowledgeable supervisor or mentor. It also seeks to include students actively in the learning process by allowing them to experiment and utilize the information they have acquired in the classroom. To enhance the management and execution of its diverse schemes and programmes, The Ministry of Information Technology and Electronics has decided to release these "Internship Guidelines," which offer a structure for the temporary employment of interns. This decision was made for the greater good of the student body.

12.4.5.2 About MeitY

The Ministry of Information and Technology (MeitY) is in charge of creating, carrying out, and reviewing national policies related to electronics, the Internet, and information technology (apart from Internet service provider licensing). Its goals are to advance e-government for the purpose of citizen empowerment; encourage the growth of the Electronics, IT, and ITeS industries in an inclusive and sustainable manner; strengthen India's position in Internet governance; and adopt a multifaceted strategy that includes human resource development, innovation and R&D promotion, efficiency enhancement through digital services, and cyberspace security.

12.4.6 NIC's Digital India Internship Programme (Summer 2023)

Through an internship, a student can gain real-world experience in the workplace while being supervised by a skilled and knowledgeable supervisor or mentor. It also seeks to include students actively in the learning process by allowing them to experiment and apply the knowledge they have learned in the classroom. At the federal, state, district, and block levels of government, the National Informatics Centre (NIC) has been instrumental in advancing the adoption of e-Government/e-Governance applications across ministries and departments. This has improved accountability and efficiency for the Indian public by fostering decentralised planning and management, increasing transparency, and improving the delivery of government services.NIC is leading the government's "Informatics-led-development" strategy, which aims to use ICT applications in public and social administration to gain a competitive edge.

NIC has released these "Internship Guidelines," which provide a framework for selecting interns for a two-month internship at the NIC headquarters in New Delhi, to improve the administration and implementation of its multiple schemes, initiatives, and projects. The purpose of these guidelines is to benefit the student community. The interns and trainees for this session will gain practical experience in designing and developing templates for user interfaces and user experiences (UI/UX) of product websites and apps.

12.4.7 R&D Plan Online Application System for Non-profit Grants

The federal government and state governments carry out several programs to promote the welfare and development of STs. VOs and NGOs, however, also contribute significantly to expanding the scope of government programs and closing vital gaps in underserved tribal areas. Providing services to remote and unreached tribal areas is the goal, and VOs and NGOs aid in this endeavor, but the caliber of those services is just as vital. Therefore, it is required of each partner VO/NGO to work towards providing high-quality services.

Through the NGO Online Application, sponsored NGOs' funding utilisation and status may be tracked, sanctions can be generated, and NGO proposals can be processed and recommended by the State, Central, and District governments. Using the online grant system, the Ministry attains openness in the funding and oversight of NGOs operating under the MoTA. Additionally, it enables NGOs to complete the necessary application processes, monitor the progress of their applications, provide completion certificates, and file grievances as needed.

12.4.8 BPO Promotion Schemes

Young job opportunities are being created by the North East BPO Promotion Scheme (NEBPS) and the India BPO Promotion Scheme (IBPS) in conjunction with the growth of the IT/ITeS business, particularly through the construction of BPO/ITeS operations in smaller cities. promoting investment in the IT and ITeS sectors to fortify the foundation of the ITeS industry and guarantee consistent growth in the region. This scheme would ensure a substantial outlay of funds. STPI is the Nodal Agency in charge of executing the BPO Promotion Schemes. The entire budgetary cost of these programs is Rs. 543 crores. The IBPS and NEBPS, respectively, are vying for 48,300 seats for BPO/ITeS operations across the country (except the Northeast) and 5,000 seats for BPO/ITeS operations in the North Eastern Region.

12.4.9 Electronic Hardware Schemes

The Electronic Hardware Technology Park Scheme, or EHTP, was introduced by the Indian government's Ministry of Electronics and Information Technology to incentivize business owners to establish facilities for producing electronic hardware. For Indian entrepreneurs to export the electronics they manufacture, the EHTP project encourages them to establish Hardware Technology Parks nationwide. The export-oriented strategy provides various incentives and support to encourage the production and export of electronic goods and services.

12.4.10 Tribal Sub Plan (TSP) and Scheduled Caste Sub Plan (SCSP

Using science and technology, the Special Component Plan for Scheduled Castes (SCSP) and Tribal Sub Plan (TSP) Schemes empower the SC and ST population and support their sustainable means of subsistence and holistic development. In an attempt to improve the lives of SC/ST groups and incorporate them into mainstream development, this Division has supported over 500 science and technology (S&T) projects under the TSP and SCSP Schemes in multiple states during the previous 20 years. These programmes have focused on various subjects, including energy, drinking water and sanitation, microenterprise development, art and craft, post-harvest technologies, health and nutrition, engineering and related professions, training and skill development, and agriculture. Through a network of sixteen NGOs, a coordinated plan on People and Protected Areas (PPA) has been implemented to develop 30,000 tribal people in forest fringe areas, surrounding animal sanctuaries and protected areas in thirteen states, including one in Utah. The programme promotes site-specific interventions and the implementation of affordable, practical technology to enhance livelihoods and protect natural resources.

12.4.11 Centres for Science, Technology, and Innovation (STI)

Scheduled Tribe (ST) and Scheduled Caste (SC) populations are being supported by the establishment of Science, Technology, and Innovation (STI) hubs around the country. Raising living standards and creating sustainable livelihoods, these hubs support developing, enhancing, and implementing pertinent and appropriate STI techniques to support the improved growth of the SC and ST populations. Twenty STI Hubs (seven for ST and thirteen for SC) have been constructed across the country's different regions in order to improve livelihood systems and boost their resilience through S&T interventions.

12.4.12 Using Wild Rose Hip Post-Harvest Technology to Empower Himachal Women

The project that the G.B. Pant Institute of Environment and Development in the Himalayas and Jagriti NGO completed is the success story. As a result of the initiative, 24 women's self-help groups (SHGs), including over 200 women, were formed. Rosehip was the subject of six workshops/training programs, protocols and procedures for sustainable harvesting were developed, two varieties of rosehip tea (rosehip and rosehip mint) were created, and technologies for rosehip seed oil were developed and tested at CSIR, IHBT, and IIT New Delhi.

12.4.13 Rural Women Technology Park, Satara, Maharashtra

With assistance from the Department of Science and Technology's SEED Division, 60 villages from the Patan block of Satara district in Maharashtra were chosen to participate in the rural women's technology project (DST).

12.4.14 Video Enhancement & Securing Women & Child Health

a DST-supported effort Under the direction of Prof. Sujit Kumar Paul of Visva-Bharati (A Central University), West Bengal, the programme Enhancement and Securing Women and Child Health through Technological Advancement and Capacity Building is supported by SEED, DST.

12.4.15 Fostering the growth of female entrepreneurs in the millet food producing industry

With assistance from the DST-SEED Division, Sri Krishna College of Engineering and Technology in Coimbatore assists female entrepreneurs in the millet food production industry to advance nutritional security.

12.4.16 Trade Related Entrepreneurship Assistance and Development (TREAD) Scheme for Women

This scheme envisages women's economic empowerment by providing credit (through NGOs), training, development, and counselling extension activities related to trades, products, services, etc.

12.5 Women in Tech and Stem Workforce

Men have always dominated the tech sector, but as we approach the next decade, it is important to consider the role that women play in this field. According to a World Bank survey, less than one-third of workers worldwide are women in technology-related sectors. Women make up 15.9% of workers in engineering and architecture and 28% of all employment in mathematics and computer fields. In the US, the percentage of workers in the STEM (science, technology, engineering, and mathematics) fields is only 23%. In the European Union, women only comprise 19.1% of the information and communication technology (ICT) workforce. In 2022, the percentage of female employees at Google climbed slightly to 33.9%.

12.5.1 Global Women's Workforce

According to recent statistics, only 47% of working-age women worldwide—less than half—are employed, and the percentage is significantly lower in some nations. This indicates that there is only one employed woman for every two men. In addition, women are more likely to face discrimination based on their gender and make less money than males. Over the past ten years, female labour participation has increased in most nations, with emerging economies setting the pace. However, development has been slower in advanced economies because of a mix of different economic policies and fewer incentives for women to return to work. It is obvious that more work needs to be done to empower women even more by expanding their educational options and fostering fulfilling career prospects. This will significantly increase the likelihood that every woman may reach her full potential and make a constructive contribution to society.

12.5.2 Gaps in STEM Education Graduate Stats

Concerning the different academic subjects, the gender gap in STEM fields is particularly concerning. The proportion of women with bachelor's degrees in computer and information sciences is only 16%; women with engineering and technology degrees make up 21%; women with economics degrees make up 27%; and women with physical sciences degrees make up 38%. Women's enrolment in STEM programs has been dropping despite equal opportunities; at present, just 18% of new computer science degrees are awarded to women. The percentage of Black and Hispanic pupils is far lower, with just 6.3% of the total student body. This thus lessens their chances of breaking into the tech industry. According to data from the U.S. Bureau of Labour Statistics, employment in STEM disciplines has surged by an incredible 79% during the last three decades and is expected to grow by an additional 11% between 2020 and 2030.

12.5.3 Hiring Trends & Retention

According to survey data, prejudice in technical hiring is still an issue, according to 65% of tech recruiters. While more than 38% of the companies surveyed desire to increase their investment in talent retention, it remains a very critical problem for businesses.

The majority of white women (66.02%) working in STEM professions for the federal government were employed in these fields. By comparison, 14.58% of the population identified as Black or African American, 9.76% as Asian,

6.42% as Hispanic or Latina, 0.97% as Native American or Alaskan Native, and 0.28% as being from the Pacific Islands or Hawaii.

Studies reveal that women are 1.6 times more likely than males to be let go from their jobs due to their ongoing lack of seniority in comparison to their male co-workers.

2022 has seen a terrible wave of IT layoffs; according to analysis by the WomenTech Network, 69.2% of those let go were women. This statistic, which highlights the disproportionate effect of layoffs on women in technology, is extremely alarming.

The information gathered indicates that greater action must be taken immediately to guarantee that women in technology have more equal chances. Together, we need to build more welcoming workplaces where women may access resources and employment opportunities equitably. To help women succeed in the computer industry, programs like mentorships, scholarships, and other support networks need to be established.

Additionally, we must work to foster an atmosphere of acceptance and tolerance that values diversity and promotes candid communication at all organizational levels. By working towards these objectives, we can guarantee that everyone has an equal opportunity to succeed in the ICT sector. WomenTech Network analyzed 4912 profiles from 54 tech businesses.

12.5.4 Women in Tech Employment Gaps

Data from the U.S. Census show that women have made great strides in the STEM fields, going from making up 8% of all STEM jobs in 1970 to 28% in 2019. Despite the progress made in 2020, there is still a notable gender gap in the GAFAM (Google, Apple, Facebook, Amazon, and Microsoft) workforce, with women making up between 28% and 42% of workers. At GAFAM, the average proportion of female employees is a mere 31%. As workers go up the corporate ladder, gender differences in the workplace become more apparent.

The largest proportion of female applicants applied for junior positions in the computer industry. However, the proportion of female applicants for employment at the mid-level and senior levels declined once more.

12.5.5 Women in Tech Retention Gaps

Unfortunately, there is still a low retention rate for women in STEM fields, even if more women are pursuing STEM degrees. NCSES data show that just 38% of women majoring in computer science were employed in the industry, while 53% of males were employed in the same capacity. Similarly, only 24% of women with engineering degrees continued their jobs after graduation, while 30% of men did the same

12.5.6 Workplace Culture Issues

According to a WomenTech Network poll, 48% of female executives listed flexibility as one of the top three things to consider before joining or remaining at a company. This shows how important flexibility is to them. At 34%, male leaders are marginally less likely than female leaders to consider this element when making decisions. Female CEOs are 1.5 times more likely to have switched careers than their male colleagues because they wanted a DE&I-focused environment.

Although working remotely or in a hybrid environment may offer a brief escape from prejudice, it is by no means a sufficient replacement for genuine systemic change. While it's true that women are experiencing less microaggressions in virtual environments, this is not necessarily reason for celebration.

According to a WomenTech Network poll, women frequently encounter microaggressions at work. A little over 64% of respondents reported being talked over in meetings, 19% said they thought stereotypes were applying to them in different contexts, and 11% said they had been asked to "supply the food" in meetings.

12.6 Leadership Position Gaps for Women in Tech

When it comes to moving up from entry-level positions to senior leadership, women encounter significant obstacles. The statistics are startling: only 87 women and 82 women of colour receive the same advancement to manager for every 100 males. There won't be nearly as many female candidates available for higher-level positions in the future, making it nearly impossible for them to ever catch up in terms of representation in these sectors.

Upon observing accomplished professionals who possess an optimal work-life equilibrium, two thirds of young women aspire to progress in their careers. By utilising this to determine and pursue attainable goals, they could advance up the corporate ladder.

12.7 Venture Capital Funding for Women-Led Startups

It is extremely difficult for women-led enterprises to secure venture capital funding; in 2020, only 2.3% of investments went to these companies. Research indicates that investment for enterprises run by women is twice as likely to come from female venture capitalists as from male colleagues, supporting the theory that female entrepreneurs will have greater success if they approach female investors. However, recent research indicates that while female-led businesses might have an advantage in attracting funding from female investors, these investments also carry the potential for discriminatory treatment based on gender. For female founders, this, therefore, creates further challenges when they later look for extra funding. Put differently, receiving initial funding from a female investor may eventually make it more difficult for an entrepreneur to secure funding from other sources.

12.8 The Impact of The COVID-19 Pandemic on Women in Technology

Due to the epidemic, women have lost a record-breaking 5 million jobs since February 2020. This is a far higher percentage than that for males and shows that the gender pay gap is still there. In addition, women are traditionally expected to handle the "invisible labor" of running a household, and the epidemic, unfortunately, made things worse. In 2021, a MetLife survey found that eighty percent of women who plan to resume their careers after a break want to work in STEM fields.

12.9 How to Give Women in Technology A Future

We have identified three areas of measures to promote and maintain women's advancement after carefully examining the most recent HR and DEI standards:

Table stakes adoption is now present in more than 75% of businesses, yet it cannot produce noticeable results. Better practices are more likely to be adopted by companies with higher percentages of female and minority employees than by those that choose for minimum requirements. Even though some businesses have adopted them, emerging techniques have produced impressive outcomes.

12.10 Leading The Way in Tech are Women

In the West, there is a widespread belief that gender inequality is far worse in areas like India, even though sexism is prevalent in the computer sector in North America, the UK, and Europe. But since this perception is false, 451 Research believed it was critical to emphasise the efforts being made by the Indian IT-BPM sector to maintain its position as the industry leader in attracting women to the workforce.

In contrast to many Western countries where women's participation in IT has stagnated or declined, the number of women employed in India's IT industry has increased rapidly over the past ten years, accounting for over 30% of all employees. As per the findings of a study conducted in collaboration with the Open University of the UK and published in NASSCOM's Women and IT Scorecard - India, 46.8% of postgraduate students in India in the field of IT and computing were female. The rate in the UK is less than half of this.

Currently, women account up 34 percent of India's IT workforce, with the bulk of these professionals being under 30. In fact, the youth of India's IT workforce has greatly contributed to its explosive growth, with the nation's STEM degree rate currently approaching gender parity at 50:50. Maintaining gender diversity in middle management and leadership positions is the next challenge. Considering NASSCOM initiatives, in order to maintain the momentum towards gender parity as it moves up to the management levels, it will be interesting to see if Indian government regulations and the outstanding work being done by IT service organizations can replicate this feat at the graduate level.

12.10.1 Context

Regardless of your political affiliation, there is growing proof that more lucrative, long-lasting companies have a more diverse workforce. According to McKinsey & Company's Why Diversity Matters research, companies that rank in the top quartile for gender diversity, for example, have a 15% higher chance of generating financial returns above their respective national industry medians.

Regardless of gender, race, or ethnicity, businesses in the lowest quartile have a statistically lower chance of generating above-average financial returns than regular businesses in the data set. Put differently, these businesses are not simply not leading, but also trailing behind. Given the realities of the tech business, which include a global skills shortage in expanding digital technology disciplines, it is even more imperative that the industry has access to a bigger talent pool.

Nonetheless, in many Western countries, the percentage of women employed in the IT sector is falling. According to research called Women in Tech: The Facts by the National Centre for Women and Information Technology (NCWIT), the percentage of women working in the US IT industry has been falling over time. In this area, women made up 36% of the workforce in 1991; by 2015, that number had dropped to 24%. Women participate at significantly lower rates in the UK. According to the annual Women in IT Scorecard published by the British Computing Society, the percentage of women employed in the industry ranged between 16 to 18% between 2005 and 2015. India is trending upward while the US is trending downward.

12.10.2 Gender Parity at the Graduate Level

The trade association for the IT BPM industry, NASSCOM, launched its first diversity campaign more than 10 years ago, when the Indian IT sector employed roughly two million people. Because the industry was growing so quickly, one issue that needed to be addressed was staff attrition. Sangeeta Gupta, SVP of NASSCOM and head of the communications and strategy division, says that diversifying the workforce is one step that will assist India in achieving that.

To recognize and reward exemplary practices in areas such as flexible work schedules, returning to work after maternity leave, employee transportation, and thorough training for HR services tasks, NASSCOM launched an annual diversity event with awards, drawing inspiration from organizations such as IBM. This was the initial stage of the industry's response to gender imbalance in India.

Ushasri Tirumala is Manhattan Associates' general manager in India and a senior vice president. Our chat made it clear that she fervently advocates for women in technology. Tirumala is co-chair of the diversity initiative WIN at Manhattan Associates. She claims that women in India have a strong cultural and political urge to work in engineering, with STEM areas being seen as a natural match and a location where they may succeed. Due to longstanding laws that encourage it, women's participation in higher education and employment in the technology sector have significantly increased in India. Compared to the US and the UK, India has shown that it can draw female students to the field of ICT.

Approximately 1.3 million women work in India's IT-BPM industry, which employs close to 3.9 million people overall. Even though this figure is significantly higher than the overall percentage of women in India's workforce (24%), research indicates that over 51% of entry-level hires are female, over 25% occupy managerial jobs, and less than 1% hold C-Suite positions. As per Tirumala, "We now need supportive policies for midlife to manage work and life in a corporate context." We need to ensure that women engineers are represented in management since there are many more of them now than there used to be."

12.10.3 Family Responsibilities

In the West, it is often known that becoming a mother and a wife hinders many women's ability to advance in their careers. This is also the case in India, where the government increased the length of maternity leave for private sector employees from 12 to 26 weeks. The Maternity Benefit Amendment Act also offers work-from-home rules, non-discriminatory performance rating systems that recognize female employees' absences, and daycare centers or creches for working mothers. As more and more Indian women become visible in the workplace, the IT services companies that have propelled India's software capabilities worldwide are working hard to establish a gender-diverse workforce. HCL Technologies, one of the biggest IT firms in India, has started a second career program called iBelieve for women who are eager to return to the profession after taking a career sabbatical.

Women who enroll in the program undergo rigorous training to improve their business and technical abilities, which are essential in today's workplace. The length and nature of the training are determined by their area of interest and prior expertise. The company reports that the program is doing well. It uses self-supporting assessments to evaluate applicants for the various courses, with the promise of an employment offer from HCL upon curriculum completion. HCL already provides childcare, flexible work schedules, work-from-home options, and "pink parking" for its female employees. 99 percent of female employees who take maternity leave are now returning to work, according to the research.

The mentorship data shows that more female respondents than male respondents attribute their work progress to other people. Compared to just 62.5% of males, women were more likely to name past or current supervisors as the element assisting their careers (74.4%). Reading books or articles was more beneficial for men (66.3%). Women are also more drawn to mentoring and professional organizations outside of the workplace. Compared to 34.4% of male participants, somewhat more female respondents (43%) attribute job progress to professional associations. Diversity programs were useful to the careers of less than half of men (6.4%), compared to 13.9% of women. Looking at the results from different ethnic groups, the respondents who view these programmes as having the most value are Black/African American (21.8%) and Latino/Hispanic (16.5%).

12.10.4 Career Inhibitors

For example, more women (42.1%) than men (26.7%) claimed that prejudice has negatively impacted their career. In addition, women feel that their inability to successfully negotiate a higher income has hurt their careers. In addition, women are far more impacted than men by the responsibility of providing family care. Nearly a quarter (24.5%) of the female respondents said that taking time off for family had hurt their career, compared to just 16.5% of their male peers. On the other hand, men placed a far higher value on a lack of work skills (21.7%) than did women (12.4%).NASSCOM and the Indian government have tackled some of the secondary hurdles women in 451 Research's US study noted, particularly in India, by passing the 2017 Maternity Bill and combating workplace sexual harassment. Indian corporations are now required by law to disclose sexual harassment incidents and their remedies in their annual reports. Indian companies are creating committees and awareness efforts to solve the problem.

12.10.5 The Future

Middle-class Indian parents still expect their boys and daughters to specialise in science or engineering because these are the desired sectors to work in. Engineering is still seen as a good profession. As the Indian market expands, this demographic will have more professional alternatives, much like in the West.

Until this shift occurs, India will remain a leading source of graduates with a gender balance in STEM disciplines. The 2015 Digital India initiative of Prime Minister Narendra Modi continues to be a powerful force in the country's efforts to change societal norms and has increased interest in maths and science, especially in smaller cities. Tier 3, 4, and 5 Indian cities are the best places to find engineering talent since state TV has significantly increased interest in STEM fields among Indian children in these Wi-Fi-less areas.

India acknowledges that much more work needs to be done in the areas of racism, caste, sexual orientation, and disability, and is by no means a model of workplace diversity. In addition, as with other places, sexism in the entrepreneurial venture capital culture in India needs to be addressed, and there should be more female CEOs in the Indian tech sector.

However, it's important to highlight that by treating the sourcing of female talent seriously, India and its leading tech companies have outperformed the West in achieving gender parity at the graduate level. India's resolve to achieve more workplace inclusivity may help the country overtake the West in the next ten years to achieve a much more diverse pipeline for leaders.

12.11 Conclusion

The study focuses on how technology affects female managers' and entrepreneurs' careers and well-being. Its conclusions demonstrate that women who use digital technologies face both advantages and disadvantages. Most of the women who were interviewed had positive opinions about digital technologies and believed that they had given them wonderful opportunities to grow in their careers.

Even though most women do not have STEM backgrounds, they can benefit from the opportunities presented by digital technologies. One could even argue that women's diverse backgrounds work to their benefit since they continue their experiential learning and see technology as a tool rather than an end in and of itself, converting the language of digital technologies into the language of companies.

References

Adrian Holliday (2016). International Journal of Bias, Identity and Diversities inEducation, pp. 1-14.

Afrah, S. H., & Fabiha, S. T. (2017). Empowering Women Entrepreneurs through Information and Communication Technology (ICT): A Case Study of Bangladesh. Management, 7(1): 1-6.

Ahmad, M., & Naseer, H. (2015). Gender Bias at Workplace: Through Sticky Floor and Glass Ceiling: A Comparative Study of Private and Public Organizations of Islamabad. International Journal of Management and Business Research, 5(3): 249-260.

Aker, J., R. Boumnijel, A. McClelland, and N. Tierney (2013). How Do Electronic Transfers Compare? Evidence from a Mobile Money Cash Transfer Experiment in Niger. Tufts University Working Paper.

Amitosh (2023). Role of Technology in the future and its impact on Society, Blog, Readers. https://timesofindia.indiatimes.com/readersblog/amitosh/the-role-of-technology-in-the-future-and-its-impact-on-society-52565/

Beena,Ms,andMadhuMathur(2012).Roleofícteducationforwomenempowerment."International Journal of Economics and Research 3(3): 164-172.

Carah N., & Dobson A. (2016). Algorithmic hotness: young women's "promotion" and "reconnaissance" work via social media body images. Soc Media Soc, 2(4).

Cummings, C. and Tam O'Neil (2015). Do digital information and communicationstechnologies increase the voice and influence of women and girls. A rapid review of theevidence. Overseas Development Institute.

Deidre, H. (2007). Applying Domestication: How the Internet Found Its Place in the Home. Managing Worldwide Operations & Communications with Information Technology. In Proceedings of the 2007 Information Resources Management Association (IRMA) International Conference: 799-801. May 19-23, Vancouver, Canada.

Dixon, L. J., Correa, T., Straubhaar, J., Covarrubias, L., Graber, D., Spence, J., & Rojas, V. (2014). Gendered Space: The Digital Divide between Male and Female Users in Internet Public Access Sites. Journal of Computer-Mediated Communication, 19(4): 991-1009.

European Commission (2014). Statistical Data on Women Entrepreneurs in Europe. Brussels: European Commission.

European Union (2017). The Life of Women and Men in Europe - A Statistical Portrait - 2017 Edition. Brussels: European Union.

Ferguson, S. J. (2016). Women and Education: Qualifications, Skills and Technology. In Women in Canada: A Gender-based Statistical Report. Ottawa: Statistics Canada.

Gronlund, A., & Oun, I. (2018). In Search of Family-Friendly Careers? Professional Strategies, Work Conditions and Gender Differences in Work-Family Conflict. Community, Work & Family, 21(1): 87-105.

Harwood, S. A. (2011). The Domestication of Online Technologies by Smaller Businesses and the 'Busy Day'. Information and Organization, 21(2): 84-106. https://doi.org/10.1016/j.infoandorg.2011.03.002

Hazen, B. T., & Byrd, T. A. (2012). Toward Creating Competitive Advantage with Logistics Information Technology. International Journal of Physical Distribution & Logistics Management, 42(1): 8-35. https://doi.org/10.1108/09600031211202454

Huyer, S., & Sikoska, T. (2003). Overcoming the Gender Digital Divide: Understanding ICTs and their Potential for Empowerment of Women. INSTRAW Research Paper Series, No. 1. United Nations International Research and Training Institute for the Advancement of Women (INSTRAW).

Jewell, T. (2011). Redefining "Women's Work": Tensions Between Technology, Entrepreneurship, and Social Reproduction. Open-Source Business Resource, July 2011: 6-12.

Joshi, A., Son, J., & Roh, H. (2015). When Can Women Close the Gap? A Meta-Analytic Test of Sex Differences in Performance and Rewards. Academy of Management Journal, 58(5): 1516-1545.https://doi.org/10.5465/amj.2013.0721

Keeley, Brian, and Celine Little 2017. The State of the Worlds Children (2017). Children in a DigitalWorld. UNICEF. 3 United Nations Plaza, New York, NY 10017.

M. Dhanamalar., S. Preethi., S. Yuvashree (2020). Impact of Digitization on Women's Empowerment: A Study of Rural and Urban Regions in India. 21(05): 107-112

M. N. Rajesh (2018). Handbook of Research on Multicultural Perspectives on Gender and Aging, pp. 177-189.

Maier, S., & Nair-Reichert, U. (2007). Empowering Women Through ICT-Based Business Initiatives: Best Practices in E-Commerce/E-Retailing Projects overview. Information Technologies and International Development, Special Issue on Women's Empowermentand the Information Society, 4(2): 43-60.

Malhotra, R. (2015). Empowering Women through Digital Technology: An Indian prospective. International Journal of Business Management.

Meade, N., & Islam, T. (2006). Modelling and Forecasting the Diffusion of Innovation - A 25-Year Review. International Journal of Forecasting, 22(3): 519-545.

Michael Jeive (2016). International Journal of Bias, Identity and Diversities in Education, pp. 53-66.

Michel Candelier and Martine Kervran (2018). International Journal of Bias, Identity and Diversities in Education, pp. 10-21.

Michell, D., Szorenyi, A., Falkner, K., & Szabo, C. (2017). Broadening Participation Not Border Protection: How Universities Can Support Women in Computer Science. Journal of Higher Education Policy and Management, 39(4): 406-422.

Mora, J. (2015). Women's Greater Educational Efforts as a Consequence of Inequality. Law, Ethics and Philosophy (LEAP), 3: 107-115.

Nath, V. (2001). Empowerment and governance through information and communication technologies: women's perspective. The International Information & Library Review 33(4): 317-339.

Notermans, Catrien (2019). "Prayers of cow dung: Women sculpturing fertile environments in ruralRajasthan (India)." Religions 10(2): 71.

Pappas, M. A., Papagerasimou, Y., Drigas, A., Raftopoulos, D., & Nikolaidis, P. (2017). ICT-based Innovation and Employability for Women. International Journal of Engineering Pedagogy (IJEP), 7(2): 36-47.

Rajahonka, M. and Kaija V. (2019). Women managers and entrepreneurs and digitalization: onthe verge of a new era or a nervous breakdown? Technology Innovation Management Review 9.6

Roy, G. (2016). Impact of Mobile Communication Technology on the Work Life Balance of Working Women - A Review of Discourses. Journal of Contemporary Management Research, 10(1): 79-101.

Scuotto, V., Serravalle, F., Murray, A., & Viassone, M. (2019). The Shift Towards a Digital Business Model: A Strategic Decision for the Female Entrepreneur. In T. Tomos, N. Kumar, N. Clifton, & D. Hyams-Ssekasi (Eds.), Women Entrepreneurs and Strategic Decision Making in the Global Economy.

Shah, H., & Saurabh, P. (2015). Women Entrepreneurs in Developing Nations: Growth and Replication Strategies and Their Impact on Poverty Alleviation. Technology Innovation Management Review, 5(8): 34-43.

Sharwan (2020). All About Electronics Hardware Technology Park (EHTP) Scheme, Blog, AVS and Associates, https://camantra.com/all-about-electronics-hardware-technology-park- ehtp-scheme/

Silverstone, R., Hirsch, E., & Morley, D. (1992). Information and Communication Technologies and the Moral Economy of the Household. In R. Silverstone & E. Hirsch (Eds.), Consuming Technologies: Media and Information in Domestic Spaces: 9-17. London and New York: Routledge.

Sinell, A., Muller-Wieland, R., & Muschner, A. (2018). Gender-Specific Constraints on Academic Entrepreneurship and Engagement in Knowledge and Technology Transfer. Technology Innovation Management Review, 8(2): 15-26.

Solesvik, M. Z. (2017). Hybrid Entrepreneurship: How and Why Entrepreneurs Combine Employment with Self-Employment. Technology Innovation Management Review, 7(3): 33-41.

Sriram (2023). Top 15 Importance of Technology in Our Daily Life [Types, Impacts, Benefits], Blog, Upgrad, https://www.upgrad.com/blog/top-importance-of-technology-in-our-daily-life/.

Van Dijk, J. A. G. M. (2012). The Evolution of the Digital Divide - The Digital Divide turns to Inequality of Skills and Usage. In J. Bus, M. Crompton, M. Hildebrandt, & G. Metakides (Eds.), Digital Enlightenment Yearbook pp. 57-75. Amsterdam: IOS Press.

Waaijer, C. J. F., Sonneveld, H., Buitendijk. S. E., van Bochove, C. A., & van der Weijden, I. C. M. (2016). The Role of Gender in the Employment, Career Perception and Research Performance of Recent PhD Graduates from Dutch Universities. PLoS ONE, 11(10): 1-16.

West, J., & Bogers, M. (2014). Leveraging External Sources of Innovation: A Review of Research on Open Innovation. Journal of Product Innovation Management, 31(4): 814-831.

Women, Connected (2019). "The mobile gender gap report 2018." GSMA, London Retrieved from https://www. gsmaintelligence. com/research.

13

Beyond Bridging the Divide Towards A Gender-Just Digital Future

Shanila S.

Department of Agricultural Extension, College of Agriculture, Kerala Agricultural University, Velanikkara, Thrissur, Kerala, India

---

Abstract

In the era of unparalleled technological progress, the digital world has become a vital aspect of everyday existence, influencing communication, economies, and information access. But even in the midst of this revolutionary trip towards a digital future, a crucial problem—the gender digital divide— remains. Beyond simple access to technology, differences in how men and women use and benefit from digital technologies result from a complex interaction of sociocultural, economic, and institutional factors. This chapter, "Beyond Bridging the Divide: Towards a Gender-Just Digital Future," examines methods for promoting a more equitable and inclusive digital environment while delving into the subtle aspects of the gender digital divide.

Keywords: *Digital environment, Technology, Environment, Communication.*

13.1 Introduction

13.1.1 Bridging the Gap: Towards a More Equitable Digital Landscape

In an age dominated by digital technologies, the concept of a more equitable digital landscape is paramount to ensuring that the benefits of technology are accessible to all segments of society. The term "digital landscape" encompasses not only the physical infrastructure, such as internet connectivity and hardware, but also the socio-economic and educational factors that influence individuals' ability to participate in the digital realm. The digital divide, often exacerbated by disparities in access to technology, internet connectivity, and digital literacy, presents a significant challenge. This paper, titled "Bridging the Gap: Towards a More Equitable Digital Landscape," seeks to explore the multifaceted nature of the digital divide and propose strategies to overcome these challenges.

Access to technology and internet connectivity stands as a cornerstone in addressing the digital divide. Research has consistently shown that individuals with limited access to devices and reliable internet face increased challenges in participating fully in the digital age (Van Dijk, 2005). Furthermore, digital literacy, encompassing the skills and knowledge required to navigate the digital landscape, is crucial. Initiatives that focus on educational programs to enhance digital literacy have demonstrated the potential to empower individuals, particularly in underserved communities (Warschauer, 2003). This chapter advocates for infrastructure development, including affordable access to devices and widespread high-speed internet, to create an inclusive digital ecosystem.

Public-private partnerships emerge as a key strategy in pursuing a more equitable digital landscape. By fostering collaboration between governmental bodies, private corporations, and non-profit organizations, initiatives can be developed to address the digital divide comprehensively. Successful case studies from around the world serve as valuable examples of the positive impact of such partnerships. For instance, projects that focus on community engagement, tailoring interventions to the specific needs of diverse communities, have shown promising results (DiMaggio & Hargittai, 2001). This collaborative approach not only contributes to infrastructure development but also enables the implementation of educational and outreach programs.

In conclusion, the quest for a more equitable digital landscape requires a holistic understanding of the factors contributing to the digital divide. By addressing issues of access, connectivity, literacy, and fostering collaborative efforts, society can work towards ensuring that the benefits of technology are shared inclusively. This chapter aims to contribute to the ongoing discourse on digital inclusion, offering insights and strategies that can guide policymakers, educators, and community leaders in their efforts to bridge the digital gap and create a more equitable future.

13.2 From Code to Culture: Decoding the Hidden Messages in Our Digital Lives

In the digital age, the intersection of technology and culture plays a pivotal role in shaping our interactions, values, and societal norms. The title "From Code to Culture: Decoding the Hidden Messages in Our Digital Lives" underscores the importance of critically examining the cultural messages embedded in technology and their impact on perpetuating social, economic, and gender divides. This part delves into the intricate web of factors, including social expectations, economic structures, and technological design, that contribute to the digital divide, focusing on how hidden cultural messages within technology can reinforce gender stereotypes.

Understanding the socio-cultural factors at play is essential in comprehending the broader implications of the digital divide. Research indicates that cultural expectations and societal norms shape individuals' perceptions of technology and influence their engagement with digital platforms (Buchi, Fieseler, & Suphan, 2016). By exploring the hidden cultural messages encoded in technology, this chapter aims to shed light on how these messages contribute to perpetuating gender stereotypes. Unpacking the ways in which technology is designed, marketed, and utilized reveals the underlying biases that can reinforce existing inequalities.

Economic structures also play a significant role in the digital divide, influencing access to technology and opportunities for participation. Gender disparities in the workplace and wage gaps can exacerbate existing inequalities, limiting women's access to educational and professional opportunities in the technology sector (OECD, 2019). By examining the economic factors intertwined with the digital landscape, this paper seeks to address how economic structures contribute to the persistence of the gender digital divide.

Moreover, this chapter advocates for critically examining technological design, as it can either perpetuate or challenge cultural stereotypes. For instance, the lack of diversity in algorithm development can lead to biased outcomes, further reinforcing existing inequalities (Crawford & Calo, 2016). By emphasizing the need to decode hidden cultural messages within technology, the paper encourages re-evaluating design practices to promote inclusivity and diversity.

In conclusion, "From Code to Culture" calls for a comprehensive understanding of the interconnected factors contributing to the digital divide, focusing on decoding the hidden cultural messages embedded in technology. By critically examining how technology reinforces gender stereotypes and contributes to social and economic disparities, this research aims to pave the way for more inclusive, equitable, and culturally sensitive digital futures.

13.3 Intersections of Injustice: Gender, Race and the Digital World

The title "Intersections of Injustice: Gender, Race, and the Digital World" highlights the crucial need to recognize the complex interplay between gender and race in shaping women's experiences within the digital landscape. This paper seeks to delve into the intricate ways in which diverse experiences within womanhood contribute to the complexities of the digital world. By examining the intersectionality of gender and race, the research aims to uncover the unique challenges faced by women from different backgrounds and explore how these challenges shape and are shaped by the digital environment.

Understanding the intersections of gender and race is essential for unravelling the multifaceted dynamics of the digital world. Existing research underscores that women of colour often face compounded challenges in terms of access to technology, representation, and opportunities in the digital space (Noble, 2018). The experiences of Black, Indigenous, and other women of colour are shaped not only by gender-based discrimination but also by racial and ethnic biases that permeate the digital landscape. This research will critically analysed how these intersecting factors influence access, participation, and representation within the digital world.

The digital landscape, while offering opportunities for empowerment, can also serve as a platform for perpetuating stereotypes and biases. Research has highlighted the presence of gender and racial biases in algorithms and artificial intelligence systems, further marginalizing women of colour (Diakopoulos, 2016). By examining these instances and exploring how diverse experiences within womanhood intersect with digital technologies, this paper aims to shed light on the ways in which technology can either challenge or reinforce existing injustices.

Moreover, this chapter will explore initiatives and strategies that address the intersections of gender and race in the digital world. By highlighting successful projects and programs that promote inclusivity and diversity, the research aims to contribute to the ongoing discourse on creating a equitable digital environment for all women, regardless of their racial or ethnic background.

In conclusion, "Intersections of Injustice" advocates for a nuanced understanding of the digital landscape by recognizing the diverse experiences within womanhood and acknowledging the influence of race on gendered experiences in the digital world. By uncovering the complexities of these intersections, the research strives to inform policies, practices, and interventions that foster a more inclusive and just digital environment.

13.4 Education and Digital Literacy: Equipping Girls for a Digital Future

In the pursuit of a more equitable digital future, education and digital literacy become paramount, particularly in addressing the gender gap that persists in the tech industry. The title "Education and Digital Literacy: Equipping Girls for a Digital Future" underscores the importance of empowering women and girls with the skills necessary to thrive in the digital landscape. This part explores strategies to close the skill gap and enhance digital literacy, providing avenues for increased participation and success for women and girls in the digital world.

One crucial strategy involves integrating digital literacy into formal education systems. By incorporating technology and digital skills into school curricula, girls can be exposed to these concepts early. Research indicates that early exposure to coding, programming, and digital problem-solving enhances confidence and interest in technology-related fields (Margolis & Fisher, 2002). Implementing inclusive educational practices ensures that girls are not only consumers but active contributors to the digital realm.

Beyond traditional education, targeted programs and initiatives aimed at promoting digital literacy among girls are essential. These programs can take various forms, including coding camps, mentorship programs, and workshops designed to build specific technical skills. Organizations such as Girls Who Code and Code.org have been at the forefront of such initiatives, offering resources and support to bridge the gender gap in technology (Huebner, 2016). By creating spaces that specifically cater to the needs and interests of girls, these initiatives contribute to breaking down stereotypes and fostering a supportive community.

Furthermore, fostering a culture of inclusivity within the tech industry is crucial. Encouraging mentorship programs, networking events, and exposure to successful women in tech can help girls envision themselves in these roles. Addressing the lack of representation is key to dismantling gender stereotypes and creating a more welcoming environment for women and girls pursuing careers in technology (Ashcraft & Blithe, 2019).

In conclusion, "Education and Digital Literacy" advocates for a proactive approach in equipping girls with the skills needed for success in the digital future. By integrating digital literacy into formal education, implementing targeted programs, and promoting inclusivity within the tech industry, we can bridge the gender gap and empower women and girls to play a vital role in shaping the digital landscape.

13.5 Entrepreneurship & Economic Empowerment: Fostering Women-Led Innovation

In the realm of technology and entrepreneurship, fostering women-led innovation is essential for achieving gender equality and economic empowerment. The title "Entrepreneurship and Economic Empowerment: Fostering Women-Led Innovation" encapsulates the focus on supporting tech startups and businesses led by women, addressing challenges related to funding and market access. This chapter explores strategies to empower women entrepreneurs, showcase their successes as role models, and advocates for mentorship programs to propel women into leadership roles within the tech industry.

One critical challenge faced by women entrepreneurs is access to funding. Research consistently highlights the gender gap in venture capital funding, with female-led startups receiving a disproportionately lower share of investment than their male counterparts (Ewens, *et al.*, 2019). To address this disparity, initiatives that provide targeted funding and resources for women-led businesses are crucial. Government programs, private sector partnerships, and angel investors focusing on women entrepreneurs create a more level playing field (Brush, *et al.*, 2019). Also, fostering a diversity culture within venture capital firms can positively impact investment decisions, ensuring a fair evaluation of women-led ventures.

Market access poses another challenge for women entrepreneurs, particularly in male-dominated industries. Networking and mentorship programs become essential tools for navigating these challenges. The industry can inspire and empower aspiring female entrepreneurs by showcasing successful women in tech as role models. Mentorship programs, where experienced professionals guide and support emerging entrepreneurs, offer valuable insights and connections (Coleman, *et al.*, 2014). Encouraging the establishment of networks that connect women entrepreneurs with mentors, industry leaders, and potential collaborators can contribute significantly to overcoming barriers and fostering success.

Moreover, advocating for policies that promote diversity and inclusion within the entrepreneurial ecosystem is vital. Encouraging corporations and government bodies to prioritize women-led businesses in procurement and partnership opportunities can provide a substantial boost to market access. The tech industry can benefit from a broader range of innovations and perspectives by creating an environment that values diversity.

In conclusion, "Entrepreneurship and Economic Empowerment" emphasizes the importance of supporting women-led innovation to achieve gender equality and economic empowerment. By addressing funding disparities, improving market access, showcasing successful role models, and implementing mentorship programs, society can foster an environment that encourages and propels women entrepreneurs towards success.

13.6 Data Equity and Algorithmic Bias

The title "Data Equity and Algorithmic Bias" underscores the pressing need to examine how algorithms, while wielding tremendous influence in various aspects of our lives, can inadvertently perpetuate gender inequality. This chapter delves into the intricate ways in which biases within datasets and algorithmic decision-making processes contribute to disparate outcomes for different genders. By exposing these issues, the research advocates for robust bias mitigation strategies and data analysis with a gender lens to ensure equitable and fair algorithmic outcomes.

Algorithms, driven by data, have the potential to either amplify or mitigate existing societal biases. Research has shown that biases within datasets, often reflective of historical prejudices, can lead to algorithmic systems that unintentionally discriminate against certain groups, including women (Diakopoulos, 2016). Whether in hiring processes, financial systems, or criminal justice, algorithms may perpetuate gender stereotypes and contribute to the reinforcement of existing inequalities. To address this, it is imperative to critically analysed algorithms and datasets, with a specific focus on identifying and rectifying gender-based biases.

Bias mitigation strategies play a pivotal role in ensuring that algorithms do not perpetuate or exacerbate gender inequality. This involves implementing techniques such as algorithmic auditing, where the decision-making processes of algorithms are scrutinized for potential biases (Barocas & Hardt, 2019). Incorporating diverse perspectives in the design and testing phases is also crucial to identify and rectify biases that may not be immediately apparent. By actively addressing biases within algorithms, we can work towards a more equitable and inclusive digital future.

Furthermore, advocating for data analysis with a gender lens is essential in promoting data equity. This approach involves examining data from a gender-specific perspective, recognizing and addressing unique challenges faced by different genders. It is vital to ensure that datasets are not only representative but also inclusive of diverse experiences, accounting for factors such as intersectionality (D'Ignazio & Klein, 2020). By incorporating a gender lens in data analysis, we can uncover and rectify biases, fostering algorithms that contribute to a fair and just society.

In conclusion, "Data Equity and Algorithmic Bias" highlights the critical importance of scrutinizing algorithms for gender bias and advocates for strategies to mitigate these biases. By exposing and addressing gender inequality within algorithmic systems, we can pave the way for a more equitable and just digital landscape.

13.7 Taming the Troll Sphere: Combating Online Violence and Harassment

The title "Taming the Troll Sphere: Combating Online Violence and Harassment" reflects the urgency of addressing the pervasive issues of cyberbullying, hate speech, and online misogyny that plague digital platforms. This chapter proposes strategies to tackle these forms of online violence, ensuring a safer and more inclusive digital environment for all users. Cyberbullying is a significant concern that disproportionately affects women and marginalized communities. One strategy to combat this issue is to implement and enforce stringent anti-cyberbullying policies on social media platforms. These policies should clearly define what constitutes cyberbullying, establish consequences for offenders, and provide mechanisms for reporting and swift resolution. Social media companies can also invest in advanced algorithms and artificial intelligence tools designed to detect and mitigate cyberbullying in real-time (Papadamou, 2020). By prioritizing user safety, platforms can create an environment where online interactions are respectful and harassment-free.

Addressing hate speech and online misogyny requires a multi-pronged approach involving both technological solutions and community engagement. Platforms can employ content moderation algorithms targeting hate speech and misogynistic content, removing such material before it reaches a wider audience (Davidson *et al.*, 2019). Additionally, fostering a sense of community responsibility is essential. Online platforms can encourage bystander intervention, empowering users to report instances of hate speech and harassment. This collective approach creates a virtual community that actively rejects toxic behaviour, promoting a culture of respect and tolerance. Education plays a crucial role in preventing online violence and harassment. Implementing digital literacy programs that educate users about the consequences of their online actions and the impact of cyberbullying on victims can foster a culture of empathy and understanding (UNESCO, 2019). These programs can target various age groups, from school children to adults, creating a more informed and responsible online community.

Moreover, legal frameworks need to be strengthened to hold perpetrators of online violence accountable. Legislation should be adapted to address the unique challenges posed by the digital realm, ensuring that those who engage in cyberbullying, hate speech, or online misogyny face appropriate consequences. This includes collaboration between governments, law enforcement agencies, and tech companies to establish effective and enforceable laws that deter online harassment (OECD, 2019).

In conclusion, "Taming the Troll Sphere" advocates for a comprehensive strategy to combat online violence and harassment. By combining robust platform policies, technological solutions, community engagement, education, and legal measures, society can work towards creating a digital space that is safer, more inclusive, and free from the scourge of cyberbullying, hate speech, and online misogyny.

13.8 Digital Rights for All: Ensuring Inclusion and Freedom

In the era of digital transformation, the title "Digital Rights for All: Ensuring Inclusion and Freedom" underscores the imperative of securing equal access and safeguarding women's online rights and freedoms. This chapter explores strategies to promote inclusivity, dismantle barriers to access, and protect women's digital rights, fostering an environment where everyone can fully participate in the digital world.

Equal access to the digital world is a fundamental aspect of ensuring digital rights for all. Research consistently highlights that woman, particularly in developing regions, face barriers to entry, including limited access to devices and internet connectivity (World Wide Web Foundation, 2019). To address this, policymakers and stakeholders must focus on implementing infrastructure development initiatives that provide affordable and widespread access to the internet. By closing the digital divide, we can ensure that women have equal opportunities to participate in the digital space.

Safeguarding women's online rights and freedoms involves addressing issues such as online harassment, privacy violations, and discrimination. Online platforms must enforce robust policies against cyberbullying, hate speech, and gender-based discrimination, creating a safer online environment (Citron & Franks, 2014). Additionally, empowering women with digital literacy skills is essential, enabling them to navigate the digital world securely and protect their online privacy. Educational programs should not only teach technical skills but also emphasize digital citizenship and the responsible use of technology (UNESCO, 2021).

Legislation plays a crucial role in protecting women's digital rights. Governments should enact and enforce laws that specifically address online gender-based violence, ensuring that perpetrators face legal consequences for their actions. Moreover, legal frameworks must be adapted to the evolving nature of digital rights, keeping pace with technological advancements to provide effective protection (OECD, 2019).

Community engagement is another key strategy to ensure digital rights for all. Creating awareness about the importance of digital rights, encouraging dialogue, and fostering support networks can empower women to assert their rights in the digital space. Collaborative efforts between civil society, tech companies, and governments can contribute to developing inclusive policies that reflect diverse perspectives and prioritize protecting women's digital rights.

In conclusion, "Digital Rights for All" advocates for a comprehensive approach to ensuring equal access and safeguarding women's online rights and freedoms.

By addressing digital divides, enforcing platform policies, promoting digital literacy, enacting protective legislation, and fostering community engagement, society can create a digital landscape that upholds the principles of inclusion, freedom, and equal opportunity for all.

13.9 Inclusive Technology Design

The title "Inclusive Technology Design" emphasizes the importance of user-centered approaches that prioritize women's needs and experiences in pursuing a more inclusive digital landscape. This chapter advocates for the integration of inclusive design principles in technology development, focusing on accessibility, language inclusion, and the consideration of diverse perspectives to ensure technology meets the needs of all users.

User-centered design is critical in addressing the unique needs and experiences of women in technology. Research indicates that women often encounter challenges related to the design of digital products and services, ranging from biased algorithms to insufficient representation in user interfaces (Bardzell & Bardzell, 2013). Incorporating user feedback, particularly from women, throughout the design process can lead to more intuitive, user-friendly technologies that better cater to a diverse range of users.

Accessibility is a cornerstone of inclusive technology design. Ensuring digital products are accessible to individuals with various abilities is crucial for gender inclusivity. For example, women with disabilities may face additional barriers when using technology, and inclusive design principles can help overcome these challenges. By incorporating features such as voice commands, screen readers, and adjustable font sizes, technology becomes more accessible to a broader spectrum of users, including women with diverse abilities (Stephenson, 2019).

Language inclusion is another key aspect of creating technology that caters to women's needs. Research has shown that language choices in user interfaces can impact how women perceive and engage with technology (Kraut, 2006). Advocating for gender-neutral language, avoiding stereotypes, and offering language options that reflect diverse cultural perspectives contribute to creating an inclusive digital environment. This is especially important as language influences how individuals interact with and perceive technology, ensuring that it does not reinforce gender biases.

Promoting diverse representation in the development teams and decision-making processes is essential for inclusive technology design. When teams bring together individuals with different backgrounds, experiences, and perspectives, they are better equipped to identify potential biases and design products that better reflect the needs of all users, including women (Wachter-Boettcher, 2019).

In conclusion, "Inclusive Technology Design" advocates for user-centered approaches that prioritize the needs and experiences of women in technology. By incorporating accessibility features, promoting language inclusion, and fostering diverse representation in design teams, the digital landscape can evolve into a space that is truly inclusive and accessible to all.

13.10 Policy and Advocacy

In the global pursuit of digital equality, the title "Policy and Advocacy" underscores the importance of both existing policies and the need for new initiatives to address the persistent digital gender gap. This chapter examines current policies aimed at promoting gender inclusivity in the digital sphere and advocates for the development of comprehensive strategies at both national and international levels to bridge the existing divides.

At the national level, many countries have recognized the importance of addressing the digital gender gap and implemented policies promoting digital inclusion. For example, some nations have launched initiatives to improve digital literacy among women, ensuring that they have the skills and knowledge to navigate the digital landscape (ITU, 2020). Additionally, policies promoting the participation of women in STEM (Science, Technology, Engineering, and Mathematics) fields contribute to reducing gender disparities in technology-related professions (European Commission, 2013). Advocacy for reinforcing and expanding such policies remains crucial to achieving lasting change.

Internationally, organizations such as the United Nations and the International Telecommunication Union (ITU) have actively promoted policies and initiatives that address the digital gender gap globally. The ITU's "EQUALS Global Partnership" aims to bridge the digital gender divide by bringing together various stakeholders, including governments, private sector entities, and civil society, to implement evidence-based solutions (ITU, n.d.). Promoting these international collaborations is essential for fostering a coordinated, global effort to tackle the digital gender gap.

Advocacy for new initiatives is paramount to overcoming emerging challenges and gaps in existing policies. As technology evolves, new dimensions of the digital gender divide may emerge, requiring innovative and adaptable policy solutions. Policymakers must engage with technology experts, gender advocates, and diverse stakeholders to stay informed about the changing dynamics of the digital landscape. By remaining proactive and responsive, policymakers can contribute to the development of forward-looking strategies that address the evolving challenges faced by women in the digital era.

In conclusion, "Policy and Advocacy" emphasizes the dual importance of recognizing existing policies and championing new initiatives to bridge the digital gender gap. By fostering collaboration between nations, organizations, and individuals, policymakers and advocates can contribute to creating a digital future that is inclusive, equitable, and accessible to all.

13.11 Digital Rights and Citizenship

In the age of digital interconnectedness, the title "Digital Rights and Citizenship" underscores the critical need for policies and practices that secure equal access to the digital world while safeguarding the online rights and freedoms of women. This chapter delves into the importance of fostering a sense of digital citizenship that promotes inclusivity, protects women from online threats, and ensures active digital participation.

Equal access to the digital world is a fundamental aspect of digital rights, as it lays the foundation for individuals, irrespective of gender, to fully participate in the benefits of the digital era. Existing disparities in access, often termed the digital divide, disproportionately affect women, particularly in developing regions (ITU, 2020). Policies and initiatives that address infrastructure gaps, promote affordable internet access, and provide digital literacy programs are essential in dismantling these barriers and ensuring women have equal opportunities to engage in the digital sphere.

Safeguarding women's online rights and freedoms requires a multifaceted approach that tackles issues such as online harassment, privacy violations, and discriminatory practices. Comprehensive legislation is crucial to deter and penalize cyberbullying and online gender-based violence. Policymakers must work towards creating an environment where women feel safe expressing themselves online without fear of harassment or discrimination (Citron & Franks, 2014). Moreover, data protection laws and privacy measures must be strengthened to shield women from unwarranted invasions of their online privacy.

Promoting digital citizenship is integral to creating a responsible and inclusive online community. Educational programs that foster digital literacy and responsible online behaviour are crucial components of digital citizenship initiatives. These programs should emphasize the ethical use of technology, respectful online communication, and an understanding of the potential risks associated with the digital world (UNESCO, 2021). By empowering women with the knowledge and skills to navigate the digital landscape safely, digital citizenship programs create a more inclusive and secure online environment.

Additionally, collaboration between governments, tech companies, and civil society is vital for effectively implementing digital rights and citizenship initiatives. Partnerships facilitate the sharing of best practices, the development of standardized guidelines, and the creation of mechanisms for reporting and addressing online abuses. These collaborative efforts can contribute to a global framework that upholds digital rights as a cornerstone of modern citizenship.

In conclusion, "Digital Rights and Citizenship" advocates for policies and practices that ensure equal access to the digital world while safeguarding the online rights and freedoms of women. By addressing the digital divide, enacting protective legislation, promoting digital citizenship, and fostering collaborative efforts, society can create an online environment that is inclusive, empowering, and respectful of individuals' rights.

13.12 The Power of Collective Action

The title "The Power of Collective Action" emphasizes the instrumental role played by women's movements and civil society organizations in driving change and advocating for digital justice. This part explores the transformative impact of collective efforts in promoting women's representation in tech leadership and fostering inclusive workplaces within the digital landscape.

Women's movements and civil society organizations have historically been at the forefront of advocating for gender equality and justice. In the digital era, these movements are leveraging their collective power to address the gender disparities pervasive in the tech industry. Advocacy initiatives spearheaded by organizations such as Women Who Code, AnitaB.org, and the National Centre for Women & Information Technology (NCWIT) focus on increasing women's representation in technology fields by providing support, resources, and networking opportunities (Ashcraft & Blithe, 2019). These organizations amplify women's voices in tech through collective action, challenging systemic barriers and working towards a more inclusive industry.

Digital justice goes beyond mere representation; it involves creating inclusive workplaces that foster equality, respect, and opportunities for everyone. Women's movements and civil society organizations advocate for workplace policies that address gender biases, promote diversity, and ensure equal pay. Initiatives such as the #MeToo movement have shed light on harassment and discrimination within the tech industry, prompting conversations and actions to create safer and more inclusive work environments (Hunt *et al.*, 2018). The collective efforts of these movements are instrumental in holding companies accountable for fostering cultures that prioritize equity and justice.

Promoting women's representation in tech leadership is key to achieving digital justice. Collective action is essential for advocating for policies that break down barriers to leadership positions. Organizations like the Geena Davis

Institute on Gender in Media focus on challenging stereotypes and biases in media, including the portrayal of women in leadership roles (Geena Davis Institute, n.d.). By collectively challenging and reshaping societal perceptions, these organizations contribute to creating a pipeline of diverse and empowered women leaders in the tech sector.

In conclusion, "The Power of Collective Action" underscores the impact of women's movements and civil society organizations in advocating for digital justice. By amplifying women's voices in technology, addressing workplace inequalities, and promoting women's representation in tech leadership, these collective efforts contribute to reshaping the digital landscape into inclusive, just, and empowering for all.

13.13 Emerging Technologies and the Future of Work

The advent of emerging technologies, including artificial intelligence (AI) and automation, has profound implications for the future of work. These technologies simultaneously offer opportunities for innovation and pose challenges to gender equity. This paper explores the impact of these technologies on the digital gender gap and proposes strategies to ensure that the future of work is characterized by equitable outcomes for all.

Integrating AI and automation in the workplace can reshape industries, creating new job opportunities and transforming existing roles. However, there is a risk that these technologies may exacerbate existing gender inequalities if not implemented with a deliberate focus on equity. Research has indicated that biases in training data can lead to gender-based disparities in algorithmic decision-making, affecting hiring processes, promotions, and performance evaluations (Crawford & Calo, 2016). To address these challenges, it is crucial to implement strategies that actively mitigate biases in AI systems and ensure fair and equitable outcomes for all employees. In the face of emerging technologies, strategies for equitable outcomes include promoting diversity in tech development and AI research. Diverse teams are more likely to identify and rectify biases in algorithms and technology applications (Buolamwini & Gebru, 2018). Encouraging inclusivity in these fields ensures that the development and deployment of emerging technologies consider a wide range of perspectives, ultimately leading to more equitable outcomes in the workplace.

Additionally, reskilling, and upskilling programs must be designed with a gender-inclusive approach. As automation transforms certain job functions, reskilling initiatives should actively engage women in gaining the skills needed for emerging roles. Governments, educational institutions, and employers must collaborate to provide accessible and gender-sensitive training opportunities that empower women to participate fully in the evolving job market (OECD, 2020).

Fostering a culture of continuous learning is vital for ensuring that women are not left behind in the fast-paced evolution of technology. Organizations should implement policies that support ongoing skill development and provide pathways for career advancement, thereby promoting a workplace environment that values and rewards diverse talents and contributions (World Economic Forum, 2020).

14.14 Conclusion

The impact of emerging technologies on the digital gender gap in the future of work is a complex challenge that requires proactive and strategic responses. By actively addressing biases, promoting diversity in tech development, implementing inclusive reskilling programs, and fostering a culture of continuous learning, society can harness the potential of emerging technologies for equitable outcomes in the workplace.

References

Ashcraft, C., & Blithe, S. (2019). "Women in IT: The Facts." National Center for Women & Information Technology (NCWIT).

Bardzell, S., & Bardzell, J. (2013). "Towards a Feminist HCI Methodology: Social Science, Feminism, and HCI." In Proceedings of the SIGCHI Conference on Human Factors in Computing Systems (CHI '13), 675-684.

Barocas, S., & Hardt, M. (2019). "Fairness and Abstraction in Sociotechnical Systems." ACM Transactions on Computation Logic, 20(4): 1-36.

Brush, C. G., et al. (2019). "Women Entrepreneurs and Venture Capital: Managing the Shadow Negotiation." Journal of Business Venturing Insights, 12: e00135.

Buchi, M., Fieseler, C., & Suphan, A. (2016). "The Dark Side of Social Media: A Literature Review." Journal of Interactive Marketing, 34: 60-72.

Buolamwini, J., & Gebru, T. (2018). "Gender Shades: Intersectional Accuracy Disparities in Commercial Gender Classification." Proceedings of the 1st Conference on Fairness, Accountability, and Transparency, 81-91.

Citron, D. K., & Franks, M. A. (2014). "Criminalizing Revenge Porn." Wake Forest Law Review, 49(2): 345-391.

Coleman, S., et al. (2014). "Tapping into Diversity: Toward an Inclusive Model of Entrepreneurial Ecosystems." Entrepreneurship Theory and Practice, 38(4): 803-833.

Crawford, K., & Calo, R. (2016). "There is a Blind Spot in AI Research." Nature, 538(7625): 311-313.

D'Ignazio, C., & Klein, L. F. (2020). "Data Feminism." MIT Press.

Davidson, T., et al. (2019). "Automated Hate Speech Detection and the Problem of Offensive Language." arXiv preprint arXiv:1908.03768.

Diakopoulos, N. (2016). "Accountability in Algorithmic Decision Making." Communications of the ACM, 59(2): 56-62.

DiMaggio, P., & Hargittai, E. (2001). From the 'digital divide' to 'digital inequality': Studying Internet use as penetration increases. Princeton University Center for Arts and Cultural Policy Studies.

European Commission. (2013). "She Figures 2012: Gender in Research and Innovation."https://ec.europa.eu/research/swafs/pdf/pub_gender_equality/she_figures_2012_en.pdf

Ewens, M., et al. (2019). "The Gender Gap in Startup Success: The Role of Accelerators." NBER Working Paper No. 26560.

Geena Davis Institute on Gender in Media. (n.d.). "Research and Resources." Retrieved from https://seejane.org/research-informs-empowers/research/

Huebner, E. S. (2016). "Teaching Girls in the Digital Age: Exploring the Impact of Female-Oriented Computing Education." Journal of Women and Minorities in Science and Engineering, 22(3): 237-257.

Hunt, V., et al. (2018). "Delivering through Diversity." McKinsey & Company.

International Telecommunication Union (ITU). (2020). "Measuring Digital Development: Facts and Figures 2020." Retrieved from https://www.itu.int/en/ITU-D/Statistics/Pages/facts/default.aspx

International Telecommunication Union (ITU). (n.d.). "EQUALS Global Partnership." Retrieved from https://www.equals.org/

Kraut, R. E. (2006). "Internet Paradox Revisited." Journal of Social Issues, 62(1): 177-196.

Margolis, J., & Fisher, A. (2002). Unlocking the Clubhouse: Women in Computing. MIT Press.

Noble, S. U. (2018). "Algorithms of Oppression: How Search Engines Reinforce Racism." NYU Press.

OECD (2019). "Bridging the Digital Gender Divide: Include, Upskill, Innovate." OECD Publishing, Paris.

OECD. (2020). "Women in the Digital Era: Breaking Through Barriers." Retrieved from https://www.oecd.org/going-digital/women-in-the-digital-era-2020.pdf

Papadamou, K. (2020). "Machine Learning Against Cyberbullying and Unwanted Exposure in Online Social Networks." Future Internet, 12(4): 68.

Stephenson, P. (2019). "Inclusive Design for a Digital World: Designing with Accessibility in Mind." Springer.

UNESCO. (2019). "Digital Literacy: A Prerequisite for Competent Digital Citizenship." Retrieved from https://unesdoc.unesco.org/ark:/48223/pf0000370384.

UNESCO. (2021). "Global Education Monitoring Report 2020: Inclusion and education: Retrieved from https://unesdoc.unesco.org/ark:/48223/pf0000377606

Van Dijk, J. A. (2005). The deepening divide: Inequality in the information society. Sage Publications.

Wachter-Boettcher, S. (2019). "Technically Wrong: Sexist Apps, Biased Algorithms, and Other Threats of Toxic Tech." W. W. Norton & Company.

Warschauer, M. (2003). Technology and social inclusion: Rethinking the digital divide. MIT Press.

World Economic Forum. (2020). "The Future of Jobs Report 2020." Retrieved from http://www3.weforum.org/docs/WEF_Future_of_Jobs_2020.pd

14

Case Studies in Women Empowerment

***Ramesh Chand Bunkar*[1*]*, Laksheeta Chauhan*[2]*, Aman Verma*[3]* and M. Vamshi*[1]

[1]*Division of Dairy Extension, National Dairy Research Institute, Karnal Haryana, India*

[2]*Rajasthan College of Agriculture, Maharana Pratap University of Agriculture and Technology, Udaipur, Rajasthan, India*

[3]*Department of Extension Education, Acharya Narendra Deva University of Agriculture & Technology, Kumarganj, Ayodhya, Uttar Pradesh, India*

Abstract

There are various ways to define women's empowerment, such as valuing women's perspectives, actively seeking them out, and elevating women's status via training, education, awareness and literacy. Women's empowerment gives them the tools and freedom to choose how to respond to many societal issues that affect their lives. They might be able to redefine responsibilities related to gender or other categories, giving them more flexibility to pursue their objectives. The empowerment of women has gained substantial attention in the fields of economics and development. It also enhances women's well-being and helps them manage risk. Approaches to support trivialized genders in a specific political or social setting may come from it. Moreover, the ability of women to make wise decisions in life that they were previously unable to make them empowered. There are some stages for women's empowerment, such as powerlessness, initiation, participation, adoption and leadership, also discussed in this chapter. There are many women who have made tremendous contributions to the universe in various sectors, such as entrepreneurship, politics, education, science and technology, etc. We all should support and contribute to women's empowerment for our nation's growth. Each one of us should teach our children from a very young age about gender equality and should never make them feel bad about their own gender.

Keywords: *Women, Women Empowerment, Literacy, Awareness, Training.*

14.1 Introduction

As discussed above, the development discourse on women nationally and internationally moved from Women in Development (WID) to Gender and

Development (GAD) and Women's Empowerment. This was reflected in the Government Plans and Policy Documents throughout the past six decades. Since independence, many schemes have been launched for the emancipation of women in the country.

The social welfare method was superseded by the development strategy in the Fifth Five-Year Plan (1974-1979). This acknowledged the contribution and productive role of women. With its emphasis on work, education, and health, the Sixth Five-Year Plan (1980-1985) marks a turning point in the history of women's growth. The Seventh Five-Year Plan (1985-1990) prioritized projects aimed at improving women's status. The 1993 adoption of the 73rd constitutional amendment to the Panchayat Raj Act marked a significant advancement for women's empowerment. According to this amendment, women will hold one-third of the positions of Sarpanch and Chairman of the district assembly (Zila Parishad) and block level legislatures (Samiti). Today, mobilization of women's groups, particularly in rural areas, has become a major strategy in bringing them into the public domain, developing leadership, and increasing their access to resources. Indian women have higher visibility in the national and state policies. The issue of resource allocation for women in all sectors is addressed by gender budgeting - an approach promoted by UNIFEM with its variations in different countries.

According to the latest UN figures, 1.36 billion people are predicted to live in India in 2019 (Ministry of Statistics & Programme Implementation, 2020). However, because the figures have been so surprising, discussing gender ratios becomes humiliating. According to the results of the 2020 Population Census, India's sex ratio is 108.176, or 924 females for every 1,000 males (Ministry of Statistics & Programme Implementation, 2020).

The problem gets more embarrassing when discussing women's empowerment. Due to their high percentage of low-skilled informal workers and their employment in low-productivity, low-paying jobs, women make up the most disadvantaged workforce in the labour market. Because of this, it's stated that women work in extremely insecure occupations for very little pay-piece rates for the most part. Deloitt in 2019 conducted a project report titled "Empowering Women and Girls in India," he said that over 195 million women, or 95% of the workforce, are either unpaid or work in the unorganized sector.

14.2 Case Study

A case study investigation entails a thorough and in-depth examination of a specific event, circumstance, institution, or social group. A case is usually described as "a phenomenon of some sort in a bounded context," with respect to both space and time. (Miles, Huberman & Saldana, 2014).

14.3 Empowerment

The crucial word in the term "empowerment" is "power." According to the International Encyclopaedia, power is defined as the ability and resources to guide an individual's life towards specific social, political, and economic objectives or standing. According to Webster's New World Dictionary, 'Empower' implies to make or cause power. Empowerment, then, is the process of gaining the ability to recognize one's rights and carry out one's obligations as efficiently as possible.

At the International Women's Conference in Nairobi in 1985, the idea of empowerment was first introduced to us. Empowerment was characterized during the conference as a shift of resources and social power in favor of women. Women can realize their potential and powers in all areas of life through the dynamic, multifaceted empowerment process. Power is not something that can be exchanged. Rather, it needs to be obtained. In addition, once obtained, it must be maintained, exercised and kept.

The idea of empowerment was first introduced to us at the International Women's Conference in Nairobi in 1985. During the conference, empowerment was characterised as a shift of resources and social power in favour of women. Through the dynamic, multifaceted process of empowerment, women can realise their potential and powers in all areas of life. Power is not something that can be exchanged. Rather, it needs to be obtained. In addition, once obtained, it must be maintained, exercised, and kept.

Empowerment is a combination of a person's skills and abilities, current support network, and fearless and adaptable approach to social policies and changes (Rappaport, 1981). According to Perkins and Zimmerman (1995), empowerment theory describes empowerment as a process and a result for communities and organisations. According to Narayan-Parker (2005), the World Bank defines empowerment as the "expansion of freedom of choice and action to shape one's life."

14.3.1 Women Empowerment

"The process of increasing women's capacity to make choices and to transform those choices into desired actions and outcomes" is one definition of women empowerment (Kabeer, 2001). This concept emphasises how crucial it is to provide women the autonomy they deserve and the ability to actively engage in the decision-making processes that impact them.

According to Bali Swain (2006), women engage in this process of challenging prevailing conventions and cultural practices in order to improve their lives and become financially independent. Through their entrepreneurial endeavours, women contribute to the generation of economic riches and innovation (Agarwal and Lenka, 2015). With creativity, a woman can express her abilities and talents and realise her ambitions when empowered (Velthouse, 1990). The Indian government has also started a number of laws and initiatives to support women's entrepreneurship, giving them more social and financial power (Agarwal and Lenka, 2018). Still, certain gaps highlight the need for increased attention to female business and empowerment.

14.3.2 Importance of Empowerment

- **i.** Boost women's self-confidence and self-esteem.
- **ii.** Promote a positive perception of women by highlighting their contributions to the business and society.
- **iii.** Help them acquire the capacity for critical thought.
- **iv.** Promote activity and decision-making via a group process.
- **v.** Give the participants the necessary information to make wise decisions about their health, work, and education.
- **vi.** Assure equitable involvement in the process of development.
- **vii.** It has the effect of supplying data, expertise, and abilities needed for financial independence.
- **viii.** Provide them with better access to legal literacy and information about their rights and entitlements so they can participate in all areas equally.

14.3.3 Status of Women in Agriculture

- Percentage of self-employed women farmers in India are- **48%**
- Employs of agriculture sector economically active women in the country **4/5th**
- With the release of Mahila Kisaan Sasakthikaran Pariyojna in 2010, **11** farm women were acknowledged as Kisaan for the first time.
- Women participation as labour force in crop farming is-**35%**
- Percentage of farm livelihoods are female headed- **20%**
- According to 2011 census **62.4%** of workforce in agriculture is women
- Less than **10%** of landowners in Thailand and India are female.
- According to a recent FAO survey, women possess only **2%** of the land, receive **1%** of all agricultural financing, and have access to **5%** of all agricultural extension resources.
- Percentage of women extension workers in world- **15%**

- According to Food and Agriculture Report 2011-2012, If women had equal access to training programs and productive resources as males, they might enhance agricultural output in poor nations by **2.5-4%** and increase farm yields by **20-30%.**
- Percentage of women agricultural scientists in the world- **6.6%**
- According to FAO percentage of female enrolment in agricultural education institutions is- **25%**
- Up to **84%** of women in rural India rely on agriculture as their primary source of income.

14.3.4 Different Enrollment Statistics

- Higher education enrollment rises to 4.14 crore, surpassing the 4-crore threshold for the first time; gains of 21% from 2014-15 and 7.5% from 2019-20 is recorded.
- The number of female registrants has reached 2 crores, up $13,000 from 2019-20.
- A notable rise in enrollment from 2014-15 of 28% for SC students and 38% for female SC students in 2020-21.
- A notable rise in ST student enrollment in 2020-21 of 47% and female ST student enrollment of 63.4%, respectively, over 2014-15.
- There was a notable rise in OBC student enrollment for female students from 2014 to 2015, of 32% and 39%, respectively.
- A noteworthy rise in student enrollment in the Northeastern Region from 2014-15 to 2020-21 of 29% and 34%, respectively. (AISHE, 2023)

14.3.5 Stages of Women Empowerment

The process of empowerment can be broken down into five steps. Powerlessness and initiation, the first two stages, are essentially passive. Here, women do not stand up for their demands and rights since they are unaware of their unfairness. At these phases, women are only beginning to realise that they can improve their own and their families' lives. During these two stages, women participate in organizational initiatives and interact in a passive manner.

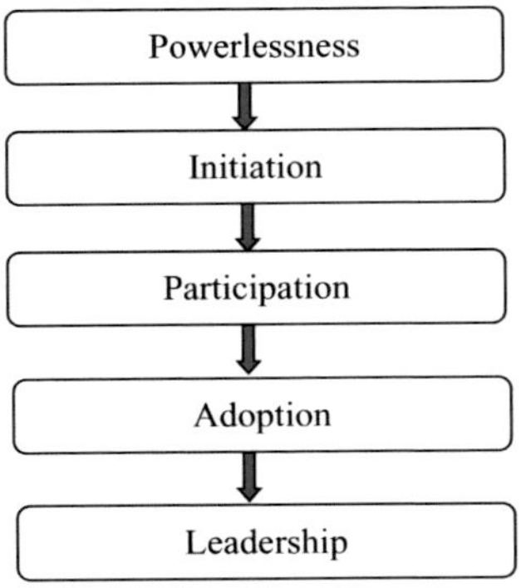

Fig. 14.1: Stages of women empowerment

The last three phases of development are seen as active phases and include leadership, adoption, and engagement. In this instance, women actively participate in the community and actively seek out interactions with local agents, organizational interventions, and other involved women. During these developmental stages, women want to know how to improve their own and their families' quality of life. They also discover where to look for resources and assistance. Ultimately, they even share their good fortune with the larger community.

14.4 Case Studies Based on Different Aspects

i. Entrepreneurship and economic empowerment
ii. Educational and skill development
iii. Political empowerment

14.4.1 Entrepreneurship and Economic Empowerment

Entrepreneurship empowerment refers to the process of enabling individuals, particularly women and marginalized groups, to develop entrepreneurial skills, access resources and create opportunities for economic self-reliance and social advancement. It involves providing support, education, and resources to individuals who aspire to start their own businesses or pursue entrepreneurial ventures, with the aim of fostering economic independence, improving livelihoods and promoting inclusive economic growth.

Eliminating the disadvantage resulting from a lack of resources is known as economic empowerment. People who are oppressed, impoverished, landless, or deprived cannot access the same resources as others who are well-off. Bringing this difference closer would put everyone on an even playing field. Moreover, a larger portion of authority over financial, human, intellectual and material resources would go to women in particular.

14.4.2 Case study: Rashmi Ratwal (jam and pickle seller)

In the training programme, Rashmi Ratwal learned how to make jam and pickles. NGO members also assisted her in obtaining funding to launch her own business. NGO members also assisted her in obtaining funding to launch her own business. With the assistance of a member of the NGO, she made an approach to STEP (Student Training for Entrepreneurial Promotion). She faced several challenges initially, such as criticism from neighbours and a lack of support from her family, but she persevered and worked hard at her task. STEP and NGO members assisted her in promoting her products by showcasing her items in several government-sponsored exhibitions such as Mahila-E-Haat and Handicraft Handloom shows. Her monthly salary rose from 2,000 to 10,000 INR in just a few months, which was a significant accomplishment for her. She uttered the following:

My husband used to beat me up a lot and spend all of his money on alcohol. In order to pay for my daily meals, I must work as a domestic helper as I lack the resources to educate my children. Following my training, I began making pickles and jams at home to sell in my community. I first used household resources and sought assistance from friends who lived nearby. Because my products are handmade, I received more orders through exhibitions. People, especially kids, like my jams, which included orange, mango and mixed fruit. Thanks to NGO and STEP, I can send my items directly to customers using a transparent supply chain strategy and without the need for middlemen. I rented a store where I set up shop properly and hired employees who are women to support me in my career. My kids are off to school, and I feel supported and happy since my husband values me too.

14.4.3 Educational and Skill Development

Education levels the playing field for all. It enables young girls to learn about their responsibilities and rights. Furthermore, acquiring information can help girls become more self-sufficient and confident in themselves. They can also benefit from advancing their intellectual, religious, political, and social consciousness. Additionally, it can stifle the spread of prejudice, intolerance, superstition, and narrow-mindedness. Gender empowerment is therefore impossible without adequate education for all.

14.4.4 Case study: From Refugee to Nobel Peace Prize Malala Yousafzai

Malala Yousafzai discovered early on that education was essential to her aspirations of becoming a doctor, freedom, and equality. While she lived in Swat, Pakistan, Malala attended a private school run by her father.

Horrifyingly, the Taliban's persecution was a harsh application of Sharia law.

As the Taliban's influence in Pakistan increased, they eventually took over the city and imposed a brutal regime that included kidnappings, public executions and the forced closure of schools, compelling Malala and the populace to escape for their safety. Her aspirations to pursue an education and life as they knew it were vanishing. Her pleas for assistance would not go unheard, as the Pakistani media gave her prominent coverage. Malala persistently advocated for education and protection through speeches, interviews on television, her blog and even meetings with American advisors. Thus, Malala was engaged in combat with the Taliban on the front lines alongside the soldiers.

The Taliban started to perceive her as a greater danger to their beliefs and their hate of the west as her popularity increased both inside and outside of Pakistan. Late in 2012, the young Malala received death threats. She was shot in the head and was in critical condition while riding a school bus to school. She was flown to the UK while medical professionals battled to preserve her life among the intricate sequence of events.

She uses nonviolence as a method and travels the world speaking to audiences, international leaders, and the UN, advocating for clear and equal access to education for all. Despite the Taliban's continued claims that she poses a threat, Malala would become the youngest person to win the Noble Peace Prize.

14.5 Political Empowerment

Political empowerment means letting women play a more significant role in government. It helps significantly in bringing women's rights to public attention. It also usurps power from those who are oppressed, disadvantaged, and without it. It also makes it possible for programs and policies to be implemented in order to assist these marginalized communities.

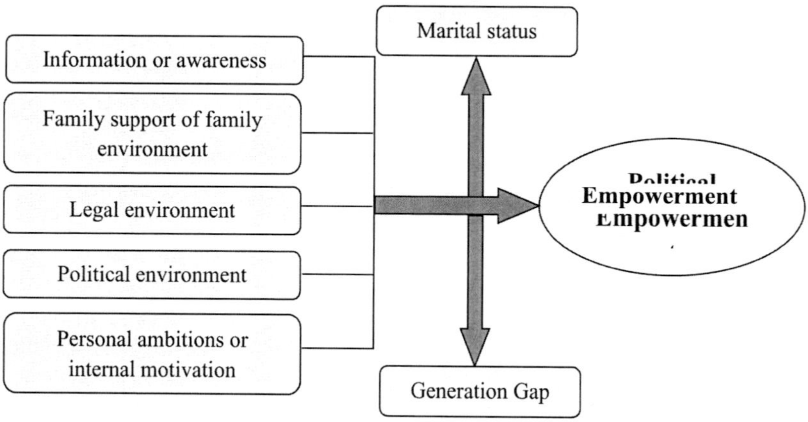

Fig. 14.2: Political empowerment model for Indian women (Eliza Sharma, 2020)

14.5.1 Case Study: Mayawati-Miracle of Democracy

Mayawati was born into a Schedule caste family on January 15, 1956, at Shrimati Sucheta Kriplani Hospital in New Delhi, with the name Kumari Mayawati Das. Her father Prabhu Das worked at the post office in Gautam Buddha Nagar's Badalpur village, which is close to Dadri. The family's daughters attended "low-performing government schools," while the males attended private schools. Mayawati completed her B.A. at the University of Delhi's Kalindi College in 1975 and her LL.B. at the esteemed Faculty of Law in 1983. In 1976, she earned her B.Ed. from the VMLG College at Meerut University in Ghaziabad. When politician Kanshi Ram paid a visit to her family's residence in 1977, while she was studying for the Indian Administrative Services exams and working as a teacher in Inderpuri JJ Colony, Delhi, he realised that she could become a leader. Mayawati was first elected to Parliament in 1989. She was the 18th Chief Minister of Uttar Pradesh, serving as such from 1995 to 1995, 1997 to 1997, 2002 to 2003, and 2007 to 2012. She was the first woman chief minister in India from a Scheduled Caste. She is the Bahujan Samaj Party's (BSP) national president. She had also represented Uttar Pradesh as a member of the Rajya Sabha from 2012 to 2017. The ascent of Mayawati from modest beginnings has been dubbed a "miracle of democracy" by former Indian Prime Minister P. V. Narasimha Rao.

14.6 Factors Affecting Women Empowerment

Many factors influence women’s empowerment. The following are some of the major factors influencing women's empowerment in India:

14.6.1 Education

Education leads to awareness about many things most importantly on health if a woman is educating enough so she can make good decisions for her and her family. She can also raise her kids well.

14.6.2 Elimination of Gender Discrimination

Eliminating discrimination and providing equal opportunities to the women at their homes as well as workstations can lead to the betterment of women’s livelihood.

14.6.3 Change of Attitude of Women

For equality women have to change their mindset they shouldn’t be dependent on any man for their basic works. Physically man and women are different and assigned role according to society are different, but it is not like to always depend on man.

14.6.4 Economic Opportunities

Economic empowerment is vital for women's autonomy and agency. Access to employment, entrepreneurship opportunities, fair wages, and financial resources allows women to support themselves economically, contribute to household income, and make independent life decisions.

14.6.5 Access to Technology

Technology, including the internet and mobile phones, can empower women by providing access to information, communication networks, education, and economic opportunities. Bridging the digital gender gap and ensuring women's access to technology and digital literacy are essential for leveraging its potential for empowerment.

14.6.6 Legal and Policy Development

Supportive legal frameworks and policies are essential for women's empowerment. Laws that promote gender equality, protect women's rights, and address issues such as discrimination, violence, and property rights are critical in ensuring women's empowerment and safeguarding their interests.

14.7 Scope For Future Women's Empowerment

To maintain gender equality and ensure women's participation and inclusion in all facets of society, more women's empowerment can be pursued in a wide range of areas. Various essential areas for additional women's empowerment encompass:

14.7.1 Economic Empowerment

Enhancing access to financial resources, promoting entrepreneurship and business ownership among women, ensuring equal pay for equal work, and addressing barriers to women's participation in the workforce.

14.7.2 Education and Skill Development

Increasing access to quality education and vocational training for women and girls, particularly in STEM fields and traditionally male-dominated industries, to expand their opportunities for career advancement and leadership roles.

14.7.3 Health and Well-being

Improving access to healthcare services, reproductive rights, and maternal health services, addressing gender-based violence and discrimination, and promoting women's mental health and well-being.

14.7.4 Political Participation

Encouraging women's participation and representation in political leadership and decision-making processes, promoting gender-sensitive policies and legislation, and addressing barriers to women's political engagement, such as discriminatory laws and cultural norms.

14.7.5 Legal Rights and Advocacy

Strengthening legal frameworks to protect women's rights, including laws against gender-based violence, discrimination, and sexual harassment, promoting access to justice for women, and supporting women-led advocacy and grassroots movements.

14.7.6 Technology and Innovation

Increasing women's access to and participation in digital technologies, closing the digital gender gap, promoting women's entrepreneurship in the technology sector, and leveraging technology for women's empowerment and social change.

14.7.7 Media Representation and Cultural Shift

Promoting gender-sensitive media representation, challenging stereotypes and harmful gender norms in media and popular culture and fostering positive social attitudes and behaviors towards gender equality.

14.7.8 Community and Social Support

Strengthening community networks and support systems for women, providing mentorship and leadership development opportunities, and creating safe spaces for women to express themselves, share experiences, and support each other.

14.7.9 Global Partnerships and Sustainable Development

Strengthening international cooperation and partnerships to advance women's rights and gender equality, integrating gender perspectives into global development agendas, and promoting sustainable development goals that prioritize women's empowerment and gender equality.

14.8 Conclusion

Women's empowerment is a dynamic and continuous process characterized by notable advancements in removing structural obstacles and creating environments that support women's growth. The shift in conversation from Women in Development (WID) to Gender and Development (GAD) to Women's Empowerment highlights the increasing awareness of the complex issues that women face. Despite progress, gender disparities persist in areas such as education, economic opportunities, and political participation, highlighting the need for continued efforts. Case studies across various domains demonstrate the transformative potential of empowering women, emphasizing the importance of holistic approaches. Collaboration between governments, civil society, communities, and individuals is essential to ensure sustained progress towards women's empowerment and a brighter future for all.

References

All India Survey on Higher Education (2023). Ministry of Education, Government of India.

Agarwal, S. and Lenka, U. (2015). "Study on work-life balance of women entrepreneurs-review and research agenda", Industrial and Commercial Training, 47(7): 356-362.

Bali Swain, R. (2006). "Microfinance and women's empowerment: evidence from the self-help group bank linkage programme in India".

Chowdary, K. R., Srishailam, B., Muralikrishnan, L. and Sravani, S. (2023). Comprehensive and Competitive Extension. Jaya publishing House, Delhi, 382.

Deloitt (2019). "Female labour force participation in India fell to 26% in 2018: Report", The Hindu, 2 May 2019.

Duflo, E. (2012). "Women empowerment and economic development", Journal of Economic Literature, 50(4): 1051-1079.

Huberman, A.M., Miles, M. and Saldana, J. (2014). Qualitative Data Analysis: A Methods Sourcebook, SAGE publications.

Kabeer, N. (2001). Conflicts over credit: Re-evaluating the empowerment potential of loans to women in rural Bangladesh. World Development, 29(1): 63-84.

Ministry of Statistics and Programme Implementation, (2020). "Population of India, 15 march 2020", available at: http://m.statisticstimes.com/demographics/population-of-india.php

Narayan-Parker, D. (Ed.) (2005); Measuring Empowerment: cross-Disciplinary Perspectives, World Bank Publications.

Perkins, D.D. and Zimmerman, M.A. (1995). "Empowerment theory, research, and application", American Journal of Community Psychology, 23(5): 569-579.

Rappaport, J. (1981). "In praise of paradox: a social policy of women empowerment over prevention", American Journal of Community Psychology, 9(1): 1-25.

Sharma, E. (2020). "Women and politics: a case study of political empowerment of Indian women", International Journal of Sociology and Social Policy, 40(7/8): 607-626.

Sunil, V. G. (2021). Fundamentals of Agricultural Extension. New Vishal Publications, New Delhi, 493-505.

Velthouse, B.A. (1990). "Creativity and empowerment: a complementary relationship", Review of Business, 12(2): 13.

15

Role of Women's in Environmental Sustainability

***Amrit Warshini*[1], *Shrestha Gautam*[2], *R.K. Doharey*[1], *N.R. Meena*[1] *and Aman Verma*[1]**

[1]*Department of Extension Education, Acharya Narendra Deva University of Agriculture and Technology Kumarganj, Ayodhya, Uttar Pradesh, India*

[2]*Department of Social Work, Jamia Milia Islamia Central University, New Delhi, India*

Abstract

Human impact on the environment is undeniable. Population, consumption habits, and technology use all play a role. We interact with the environment in three ways: as a source of materials, a place to live, and a place to dispose of waste. To address environmental issues effectively, we need a deep understanding of the environment itself. Sustainability is especially important when resources are limited. This is particularly true in places facing poverty, population growth, and unpredictable weather, where environmental pressures are even greater. Women often play a crucial role as environmental caretakers in such settings, yet their contributions are frequently overlooked. Many environmental initiatives fail because they neglect the importance of gender dynamics. Focusing solely on men as the solution has proven inadequate. To succeed, we must recognise women's vital role in maintaining a healthy environment. This requires a significant shift from the current approach, which positions women at the centre of environmental efforts. This chapter proposes a framework for acknowledging and valuing women's contributions to environmental problem-solving. It takes into account the social, cultural, and physical contexts (the "place") in which women operate when managing the environment. Ensuring environmental security for future generations is essential, especially as global environmental changes threaten sustainable development.

Keywords: *Women empowerment, Sustainable, Gender dynamics, Environment*

15.1 Introduction

15.1.1 Empowering Women for a Sustainable Future in India

Education and opportunity are key to a thriving future, and investing in women's education empowers not just them but also the next generation. India's vast natural resources offer countless benefits, but centuries of human activity have taken a toll. Unsustainable practices and mismanagement have degraded the environment, threatening the well-being of communities across the country. The evidence of environmental decline is undeniable. Over the past decade, the situation has worsened. Our actions are pushing us towards a potential environmental crisis. India, like the earth itself, sustains a multitude of activities. This has undeniably improved lives, particularly in recent decades. However, our growing disconnectedness has brought new challenges, with environmental issues becoming a major public concern.

Indiscipline in human behavior is a major root cause of environmental degradation. India faces a multitude of man-made environmental disasters, including oil spills, deforestation, desertification and air and water pollution. The link between poverty and environmental issues states that our inability to tackle problems like desertification and water scarcity stems from these fundamental challenges. The world is recognizing the gravity of the environmental crisis. There's a growing consensus that environmental problems threaten human survival and sustainable development. Uncontrolled human activities are putting immense strain on the environment's capacity to support us.

Our dependence on the environment has never been greater. We rely on it for basic needs like food, water, and shelter. However, population growth, increasing consumption, and scientific advancements have negatively affected our environment. This is particularly concerning in developing countries like India, where environmental issues are compounded by poverty, inequality, and unequal gender dynamics. A degraded environment harms everyone. The negative impacts are widespread and costly. There's a growing awareness of the need for sustainable practices. However, many environmental initiatives in India have neglected the role of women, who make up over half the population and are deeply connected to the environment.

15.1.2 Women: Key Players in India's Sustainable Development Journey

While achieving gender equality by 2030 seems far-fetched, the UN's Sustainable Development Goals (SDGs) offer a potential turning point. Gender is a crucial concept in this discussion. It refers to the social and cultural construction of masculinity and femininity, rather than solely biological differences. Despite being the backbone of rural communities in India - farmers, water carriers, and fuelwood collectors - women are often excluded from environmental sustainability projects that assume these tasks belong to men. This leads to a decline in women's access to productive land as development, industrialization, and urbanization. Furthermore, they are often denied access to education on modern technologies and innovations. Women's efforts in environmental protection are finally gaining recognition. Success stories like the Chipko movement in India demonstrate the power of women's leadership in achieving sustainability.

15.1.3 Women in Development

The concept of Women in Development (WID) emerged as a critical approach to achieving progress in developing countries, including India. A core concern for the women's movement globally, WID aimed to elevate women's status by recognizing their existing roles and integrating them more fully into the economy.

WID emphasizes enhancing women's economic capabilities and contributions. Historically, women's exclusion from the workforce was seen as a key factor in their subordinate position. WID argued that by actively participating in the economy, women would not only contribute to national development but also improve their own standing within society. WID challenges the limited view of women solely as wives and mothers. Many development policies confined women to these roles. WID asserts that women are far more - they are vital and productive members of society. They should not be seen as passive recipients of aid, but rather as active participants and economic contributors.

In essence, WID recognizes women's economic power. They are not simply a missing piece of the development puzzle, but a previously undervalued resource. WID emphasizes that women are essential for driving economic growth in India. India can unlock its full potential by integrating their talents and skills into the development process.

15.2 Major Environmental Challenges in India

Understanding the role of women in protecting the environment requires examining India's most pressing environmental issues. This section focuses on problems that are difficult to reverse once they take hold. These include deforestation, desertification, and issues related to soil, water, air, and industrialization.

Fig. 15.1: Figure shows the Contemporary Problems of Ecology

(*Source:* dreamstime.com/z/contemporary-problems-ecology-pollution-air-water-soil- natural-resource-depletion-waste-garbage-disposal-deforestation)

15.2 Summary Table of Major Environmental Challenges in India

S.No.	Environmental Challenge	Description
1	Deforestation	Cutting down trees for firewood, agriculture, and poor forest management lead to soil erosion, flooding, reduced water flow, climate change, and biodiversity loss.
2	Desertification	Overgrazing, unsustainable farming practices, and climate change exacerbate the process of fertile land degrading into arid, unproductive landscapes.
3	Soil Problems	This includes soil degradation, desiccation (drying out), and erosion, which reduce water bodies' fertility, productivity, and sedimentation.
4	Air Pollution	Emissions from industries, vehicles, and the burning of agricultural waste and cooking fuels contribute to air pollution, which can cause health risks and environmental damage.
5	Industrialisation	While promoting economic growth, industrialisation also leads to the overexploitation of natural resources, air and water pollution, and land degradation.
6	Water Pollution	Untreated industrial effluents, sewage, agricultural runoff, oil spills, and soil erosion contaminate water sources, harming public health and aquatic ecosystems.
7	Urbanisation	Rapid urbanisation strains resources like water and energy, generates waste, contributes to air pollution, and leads to the formation of slums with poor sanitation.

15.2.1 Deforestation

Indiscriminate tree cutting down for firewood, clearing land for agriculture, and poor forest management are leading to a complex set of environmental problems in India. These consequences include soil erosion, increased flooding, reduced river water flow, and changes in local climates. Deforestation also contributes to the extinction of plant and animal species, reduces the amount of carbon dioxide stored in the soil, and makes agricultural land more vulnerable to droughts.

15.2.1.1 Causes of Deforestation in India

i. **Fuelwood Collection:** Many of India's population relies on firewood for cooking and heating. Unsustainable collection practices outpace natural regeneration, leading to forest depletion.

ii. **Agriculture:** Expanding agricultural land for crops and livestock farming often involves clearing forests. While some methods (shifting cultivation) can be sustainable, current practices often lead to permanent deforestation.

iii. **Infrastructure Development:** Building roads, dams, and other infrastructure projects often necessitates clearing forests, fragmenting habitats and impacting biodiversity.

iv. **Industrial Activities:** Logging for timber, mining, and paper production contributes to deforestation, with limited reforestation efforts.

v. **Livestock Grazing:** Overgrazing by livestock can damage forest ecosystems, preventing tree regeneration and leading to soil erosion.

15.2.1.2 Impacts of Deforestation in India

i. **Soil Erosion:** Without trees to hold the soil in place, heavy rains lead to increased erosion, reducing agricultural productivity and silting rivers and reservoirs.

ii. **Droughts:** Deforestation disrupts natural water cycles, reducing rainfall and exacerbating drought conditions in some regions.

iii. **Biodiversity Loss:** Forests are home to many plant and animal life. Deforestation destroys habitats, leading to species extinction and a decline in ecosystem services.

iv. **Climate Change:** Trees absorb carbon dioxide, a greenhouse gas. Deforestation increases atmospheric carbon dioxide levels, contributing to global warming.

v. **Livelihood Loss:** Forests provide communities with resources like timber, non-timber forest products, and medicinal plants. Deforestation threatens these livelihoods and traditional ways of life.

15.2.2 Desertification

This process turns once-productive land into barren waste. It's caused by a combination of human activities and climate variations. Desertification reduces soil fertility and overall land productivity. In India, the Thar Desert is expanding southward at an alarming rate, threatening agricultural land and livelihoods. While overgrazing, excessive farming, and unsustainable use of firewood contribute to desertification, the impact of climate change cannot be ignored. Desertification is the process by which fertile land degrades into arid, unproductive landscapes. India faces this threat, particularly in the Thar Desert region.

15.2.2.1 Efforts to Combat Deforestation and Desertification

The Indian government and various organizations are taking steps to address these issues, including:

i. **Afforestation and Reforestation Programs:** Planting trees and restoring degraded forest lands are crucial for mitigating deforestation and desertification.

ii. **Promoting Sustainable Forestry Practices:** Sustainable logging techniques and community-based forest management can help conserve forests while meeting local needs.

iii. **Soil Conservation Measures:** Techniques like terracing, cover cropping, and mulching can help prevent soil erosion and improve land fertility.

iv. **Promoting Water Conservation:** Drip irrigation, rainwater harvesting, and efficient water management practices are essential for combating desertification.

v. **Raising Awareness:** Educating communities about the importance of forests and sustainable land management practices is crucial for long-term solutions.

vi. Addressing deforestation and desertification requires a multi-pronged approach, combining government initiatives, community involvement, and technological advancements. India can protect its precious environmental resources for future generations by taking decisive action.

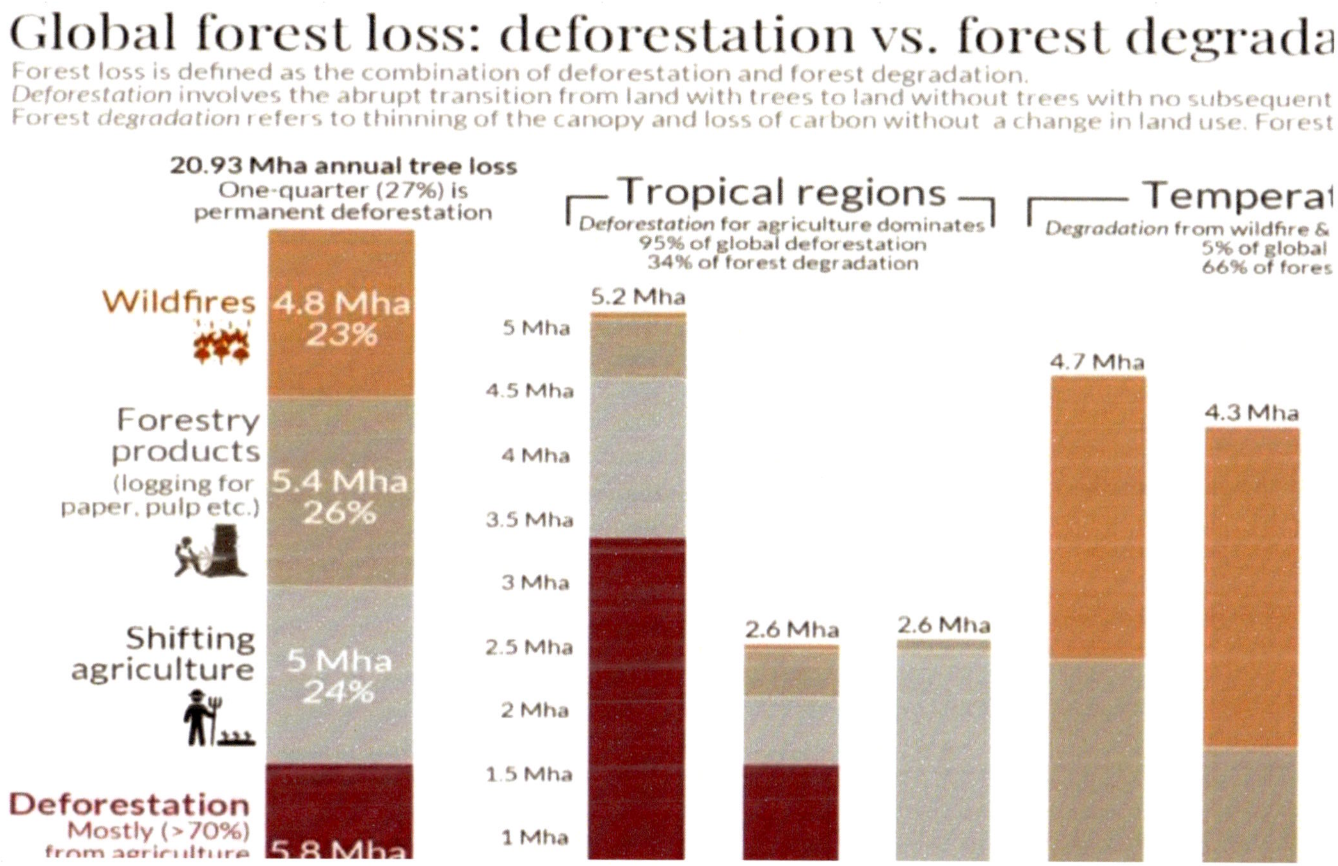

Fig. 15.2: Summary Table of Major Environmental Challenges in India.

15.2.3 Soil Problems in India: A Threat to Sustainability

Maintaining healthy soil is essential for a productive agricultural sector and a thriving environment. However, India faces several major soil-related challenges:

i. **Soil Degradation:** This refers to a decline in soil fertility and productivity. This is often caused by practices like intensive cropping without proper replenishment of nutrients, burning crop residues, and leaving soil exposed to the elements. These activities can lead to a decrease in organic matter, essential for healthy soil structure and nutrient retention.

ii. **Soil Desiccation:** This is the drying out and hardening of soil, making it unsuitable for plant growth. This can be caused by factors like deforestation, which reduces rainfall and increases evaporation. Climate change, with its erratic rainfall patterns, can also exacerbate soil desiccation. Once this process occurs, it's very difficult to reverse.

iii. **Soil Erosion:** This is the removal of topsoil by wind or water. Deforestation, overgrazing, and poor farming practices all contribute to soil erosion. Erosion not only reduces soil fertility but also leads to sedimentation in rivers and reservoirs, impacting water quality and infrastructure.

The Severity of the Problem: The intensity of these soil problems varies depending on factors like population density, land use patterns, and agricultural practices. However, the negative consequences are widespread, affecting agricultural productivity, food security, and overall environmental health.Techniques like crop rotation, cover cropping, and composting can help to maintain soil health and fertility even terracing, contour plowing, and windbreaks can help to reduce soil erosion from wind and water. Efficient irrigation methods like drip irrigation can help conserve water and prevent soil salinization.Educating farmers about the importance of sustainable soil management practices is crucial for long-term solutions.

15.2.4 Air Pollution: A Silent Killer in India

Air pollution is a severe environmental threat with far-reaching consequences for human health. While it's a global concern, some areas suffer more than others due to concentrated pollution. This is especially true in densely populated regions.

15.2.4.1 The Global Impact of Air Pollution

i. **Unequal Distribution:** Air pollution disproportionately affects developing countries. Globally, 92% of people breathe unhealthy air, primarily driven by fossil fuel use. This not only pollutes the air but also contributes to climate change (WHO, 2016).

ii. **Health Risks:** The World Health Organization (WHO) considers air pollution the greatest environmental health risk. Microscopic pollutants penetrate the respiratory and circulatory systems, increasing the risk of cancer, stroke, heart disease, and lung disease. These air-related illnesses cause an estimated 7 million premature deaths worldwide each year, with 90% occurring in low- and middle-income countries (WHO, 2018).

15.2.4.2 Air Pollution in India

India faces a significant air pollution challenge, with high levels of emissions from:

Industrial activity releases pollutants into the air, harming human health and the environment whereas vehicle emissions are a major source of air pollution, particularly in urban areas with heavy traffic congestion. Burning agricultural waste contributes to air pollution, especially during harvest seasons even indoor air pollution occur from unclean cooking fuels (like firewood and dung) poses health risks, particularly for women and children.

15.2.4.3 Addressing air pollution in India requires a multi-pronged approach

i. **Shifting to Clean Energy:** Promoting renewable energy sources like solar and wind power can help reduce reliance on fossil fuels and decrease air pollution.

ii. **Improving Vehicle Standards:** Implementing stricter emission standards for vehicles and promoting electric vehicles can significantly reduce air pollution from transportation.

iii. **Sustainable Agricultural Practices:** Encouraging practices that minimize burning of agricultural waste can help improve air quality.

iv. **Promoting Clean Cooking Solutions:** Providing access to cleaner cooking fuels and technologies can significantly improve indoor air quality, especially in rural areas.

15.2.5 Industrialization

Industrialization refers to a nation's shift from an agrarian economy to one driven by manufacturing. This process typically involves:

i. **Increased Importance of Industry:** Manufacturing becomes a dominant force in the economy, contributing significantly to Gross Domestic Product (GDP).

ii. **Rising Living Standards:** Industrialization often leads to increase per capita income and improved living standards.

iii. **Urbanization:** People move from rural areas to cities to find jobs in factories, leading to growth of urban centers.

iv. **Shift in Production:** Production moves from homes and small workshops to large-scale factories.

2.5.1 The Environmental Impact of Industrialization

While industrialization has brought economic benefits to India, it has also put a strain on the environment. Industrial activities require vast amounts of natural resources, leading to concerns about over exploitation. Factories generate significant air and water pollution through emissions and waste disposal. Land degradation also occurs due to industrial waste and improper management practices whereas industrial facilities tend to concentrate in specific zones, leading to localized environmental problems in those areas.

India needs to find ways to balance economic growth with environmental sustainability. Here are some potential solutions

i. **Promoting Clean Technologies:** Investing in cleaner production methods and renewable energy sources can reduce pollution and resource extraction.

ii. **Stricter Environmental Regulations:** Enforcing stricter environmental regulations and penalties for polluters can incentivize cleaner practices.

iii. **Sustainable Industrial Practices:** Adopting sustainable waste management and resource utilization can minimize environmental damage.

15.2.6 Water Pollution

Water pollution is a growing concern in India, posing a major threat to public health and environmental well-being. Untreated industrial effluents laden with chemicals and toxins are a significant source of water pollution in India. These contaminants can render water sources unusable for drinking, irrigation, and aquatic life whereas inadequate sewage treatment facilities lead to the release of raw sewage into rivers and other water bodies. This organic waste depletes oxygen levels in the water, harming aquatic ecosystems and increasing the risk of waterborne diseases. Excessive use of fertilizers and pesticides in agriculture can contaminate water sources through runoff. These chemicals can harm human health and disrupt aquatic life and even accidental oil spills and leaks from refineries and pipelines can devastate water quality in affected areas. The impacts can be long-lasting, affecting fisheries and recreational activities. Land clearing and construction activities can increase soil erosion leading to sedimentation in rivers and reservoirs which reduce water quality and storage capacity.

15.2.6.1 Consequences of Water Pollution

Waterborne diseases like diarrhea, cholera, and typhoid fever are prevalent in areas with contaminated water supplies. These illnesses can be particularly harmful to children and vulnerable populations. Pollution can render water sources unfit for various uses, leading to water scarcity for drinking, irrigation, and industrial purposes. Water pollution harms aquatic ecosystems, leading to a decline in fish populations and biodiversity.

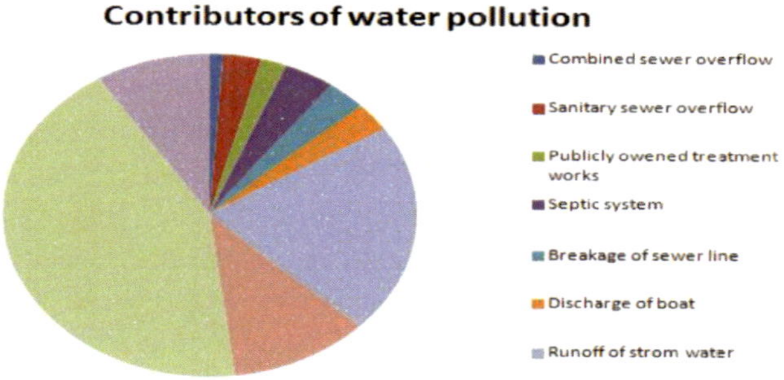

Fig. 15.3: Shows Contributors of water Pollution
(*Source*: https://www.pdfprof.com/PDF_Image.php)

15.2.7 Urbanization

Urbanization refers to the movement of people from rural areas to cities, leading to the growth of urban centers. This is a global trend, and India is experiencing rapid urbanization In 1950, only a small percentage of Indians lived in cities. Today, that number is much higher and is projected to continue increasing. The rapid growth of cities often outpaces the ability to provide adequate housing for everyone. This leads to the formation of slums, characterized by overcrowding, poor sanitation, and inadequate infrastructure. Large urban populations generate significant amounts of waste, which can be difficult to manage effectively. This can lead to pollution and health problems. Urbanization puts a strain on natural resources like water and energy. Additionally, increased traffic and industrial activity contribute to air pollution.

15.2.7.1 Opportunities in Urbanization

Cities are often hubs of economic activity, with opportunities for jobs and business development. Urban centers can foster innovation and creativity, leading to new solutions for urban challenges. Cities can provide better access to education, healthcare, and other essential services.

India needs to manage its urbanization process effectively to reap the benefits while minimizing the drawbacks. Developing cities with efficient infrastructure, proper waste management systems, and green spaces can improve the quality of life for residents whereas investing in upgrading slums with better housing, sanitation, and basic amenities can improve living conditions for vulnerable populations and by developing efficient and affordable public transportation systems can reduce traffic congestion and air pollution. Also, by utilizing smart technologies like waste management systems and energy-efficient buildings can contribute to a more sustainable urban future.

15.3 Sketches of Contribution of Women to Sustainable Environment in India: A Growing Recognition of Women's Role

Policymakers are increasingly acknowledging the gravity of environmental problems and the need for solutions. India's National Policy on Environment (1989) marked a significant shift towards environmental awareness. This policy aligns with global calls for sustainable development through responsible environmental management. Major international organizations are actively involved in addressing environmental challenges. However, a crucial aspect has been overlooked: **the vital role women can play in environmental sustainability.**Indian women are known for their resourcefulness and have traditionally been the guardians of the environment. They have borne the brunt of environmental degradation yet are rarely considered in environmental planning and implementation. Gender-balanced policies are essential, considering the root causes of environmental neglect lie in gender roles and labor allocation.

15.3.1 Women's Everyday Environmental Contributions

Throughout history, women have been associated with nature - worshipped as Earth Mother, protectors of water sources, and caretakers of biodiversity. Their daily activities directly connect them to the environment:

Women play a crucial role in managing forest resources. They collect fuelwood, fodder, and non-timber forest products, and their knowledge of sustainable harvesting practices is invaluable. The Chipko movement, a successful non-violent protest against deforestation in the Himalayas, was spearheaded by women like Gaura Devi. They are actively involved in practices like composting and using organic manure to maintain soil health and productivity. Traditionally they were responsible for fetching water for household use, they have a deep understanding of water conservation techniques and they play a key role in protecting water sources and promoting sustainable water usage. In many rural communities, women are responsible for household waste management practices like composting and segregation.

Despite their significant contributions, women's knowledge and experience are often disregarded in environmental policymaking. Large-scale irrigation projects and commercial farming practices, often driven by men, can lead to environmental degradation like salinization. Women's traditional ecological knowledge offers valuable insights for sustainable resource management.

Including women in environmental decision-making processes is crucial and by providing women with training and resources to enhance their skills in sustainable resource management. Empowering women through microfinance initiatives can enable them to adopt eco-friendly practices and participate in sustainable livelihood options. Encouraging women's active participation in village councils and environmental committees to ensure their voices are heard.

15.3.2 Framework for Recognizing Women's Contributions to a Sustainable Environment

This framework draws inspiration from the "Gender Challenge" and aims to shed light on the crucial role women play in environmental sustainability. Comprehensive databases capturing the types and geographical extent of environmental problems across India's diverse ecological zones. This data should be disaggregated by gender to accurately reflect women's contributions to environmental management and beyond traditional economic data collection methods that often overlook women's contributions, such as their work in food processing and water management.

i. **Indigenous Knowledge and Expertise:** By documenting and valuing the time-tested knowledge and skills women possess in managing their environment.This knowledge is critical for understanding sustainable practices and coping mechanisms in the face of environmental challenges.Actively incorporating women's indigenous knowledge into environmental policy and decision-making processess will ensures their perspectives are heard and valued.

ii. **Empowering Women through Participation and Capacity Building:** Environmental decision-making ensures women's active participation at all levels, from community councils to national planning bodies.

    This requires dismantling existing gender biases and promoting equal representation. Equip women with the knowledge and skills needed to address contemporary environmental challenges effectively.Designing capacity-building programs that cater to the specific needs and circumstances of women in India.

iii. **Gender-Sensitive Strategies and Resource Allocation:** Shift from "planning for the people" to "planning with the people." This ensure at least 50% of women are involved in all stages of environmental projects, from conceptualization and environmental impact assessments to implementation, monitoring, and evaluation. Directing resources towards environmental projects that actively consider and address the specific needs and contributions of women and by allocation of funding strategically to address past and promote gender equity in environmental initiatives.

iv. **Good Governance:** Good governance encourages women's active participation and visibility in environmental issues. Democratic frameworks and policies can empower women and facilitate collaboration between public and private sectors and with non-governmental organizations to achieve a more sustainable future.

15.4 Women: Stewards of the Environment

Studies consistently highlight the crucial role women play in environmental protection and preservation. Their daily lives are intricately linked to the environment, fostering a deep understanding and appreciation for its delicate balance. In developing countries, women are often directly involved in agricultural activities, collecting water and firewood, tasks that connect them intimately with nature. This close relationship fosters a sense of responsibility for environmental well-being. They understand the importance of soil health, plant life, and water resources, leading them to adopt sustainable practices and avoid wasteful exploitation.Women are vital in transmitting environmental knowledge and values to future generations. They teach children about responsible consumption, such as saving water and electricity and reducing waste. This cultural transmission ensures a more environmentally conscious future.

History is replete with examples of women who have spearheaded environmental movements.

i. **The Bishnoi Movement (1700):** Led by Amrita Devi, villagers, mostly women, hugged trees to prevent them from being cut down by the king's soldiers.

ii. **The Chipko Movement (1973):** Women like Gaura Devi and Sudhesha Devi actively participated in this movement, embracing trees to prevent deforestation in the Himalayas.

iii. **The Silent Valley Movement (1978):** Sugutha Kumari and other women played a key role in protecting the Silent Valley Forest from destruction by a hydropower project.

iv. **The Narmada Bachao Andolan (1985):** Medha Patkar, along with others, led this movement against the Sardar Sarovar Dam project, highlighting environmental concerns.

v. **Beyond Chipko:** Women have been central figures in other environmental movements, including the Appiko movement, the Silent Valley movement, and the Narmada Bachao Andolan. Medha Patkar's work with the Narmada Bachao Andolan is a prime example of women's leadership in environmental protection.

vi. Women like Sunita Narain and Vandana Shiva promote environmental sustainability through organizations and advocacy.

vii. **Maneka Gandhi:** Beyond being the wife of a prominent politician, Maneka Gandhi is a champion for both animal rights and environmental causes. In 1994, she established People for Animals, India's largest animal welfare organization. She actively promotes ahinsa (nonviolence) and believes India needs a strong movement against animal cruelty.

viii. **Vandana Shiva:** A physicist and philosopher of science, Vandana Shiva is a leading environmentalist and Eco-feminist based in Delhi. Inspired by Gandhian principles, she tirelessly works to protect forests, empower women through networks, and conserve local biodiversity. She has authored influential books and founded the Navdanya movement, a national effort to preserve biodiversity through seed banks and organic farming practices. Her dedication earned her recognition as a "Time Magazine environmental hero" in 2003.

ix. **Sugatha Kumari:** Poet and environmentalist Sugatha kumari has dedicated her life to celebrating nature through her writing. A prominent figure in environmental and feminist movements in Kerala, she founded the Prakriti Samrakshana Samithi organization. She actively participated in the "Save Silent Valley" movement, a successful campaign to protect a precious evergreen forest.

x. **Radha Bhatt:** Concerned about the environmental impact of hydro-electric projects, Radha Bhatt played a key role in the Uttarakhand Nadi Bachao Abhiyan movement launched in 2008. She led a 2,000-kilometer march to raise awareness about the threats posed to the Ganges River and its tributaries, as well as the fragile Himalayan ecosystem.

Fig. 15.4: Activities During Chipko Movement
(*source:* https://image.slidesharecdn.com/chipkomovement-)

15.5 Women on the Global Stage

The inspiring work of Wangari Maathai, founder of the Green Belt Movement in Kenya, demonstrates the power of women to lead large-scale environmental restoration efforts. By planting millions of trees, her movement addressed deforestation while empowering rural women and promoting economic growth.

15.5.1 Legal Framework and Women's Role

The Indian Constitution guarantees equal rights and environmental protection. However, women often go beyond legal frameworks and actively seek justice through the courts, as seen in cases like Krishna Gopal vs State of Madhya Pradesh and Kinkari Devi vs State of Himachal Pradesh. These cases highlight women's proactive approach to environmental management.

15.5.2 Challenges and Realities

Rapid population growth, economic pressures, and climate change pose serious threats to India's environment. Simply promoting gender equality isn't enough to solve these complex issues. Deep-rooted problems like unequal workloads, societal expectations, poverty, and conflicting priorities need to be addressed.

There's growing recognition that women play a vital role in environmental sustainability in India, similar to other parts of the world. Women's unique experiences and knowledge are crucial for developing successful environmental initiatives. Environmental projects must actively involve women at all stages, from planning to implementation and evaluation. This requires a shift beyond just empowering women, to truly integrating them as key decision-makers.

By harnessing women's expertise, we can effectively combat environmental degradation and create a more sustainable future for India.

15.6 Women as Agents of Change

Including women in environmental decision-making is essential. Men and women typically play different roles in environmental issues, soa gender-sensitive approach is crucial. Recognizing the diverse social, cultural, and ecological contexts across India is also critical.

15.7 Conclusion

The vital contributions women make to environmental sustainability in India. By integrating gender equality into environmental policies and projects, we can create a more equitable and sustainable future for all. Achieving these goals requires collaboration between communities, governments, and businesses to create a supportive framework for sustainable practices.

References

Aditya, S. K. (2016). Role of Women in Environmental Conservation. International Journal of Political Science and Development, 4(4): 140-145.

Agarwal, A., D' Monte, D. and Samarth, U. (1987a). The Wrath of Nature: The Impact of Environmental Destruction on Floods and Droughts. New Delhi Centre for Science and Environment.

Awasthi, Bela (2002). Women's Liberation and Modern World. Published by R. Kumar, Omsons Publications.

Bhartiya, Tribhuwan Kumar. Assessment of Environmental Awareness among the Public of Assam (India).

Bowonder, B. (1986). Deforestation in Developing Countries. Journal of Environmental Systems, 15: 171-192.

Dodge, Norton (1973). Women in Economic Development: A Review Essay. International Review of Education, 19(1): 161-166.

International Journal of Applied Environmental Sciences (2017). Nov; 7(12): 1359-1365. ISSN0973-6071.

Kalita, Dr. Jovial. Malik, Dr. Abdul Foyes Md. Violence against Women in India. Digboi Mahila Mahavidyalaya; c2017, 1.

Monk J. and Hanson, S. (1982). "On Not Excluding Half of the Human in Human Geography". The Professional Geographer, 32: 11-23.

Odoemena C.S.I. and Egwali E. C. (2006). Fundamentals of Conservation and Development of Natural Resources in Nigeria. Unwana Graphics Press. ISBN: 978-38584-1-6.

Pathak, Guptajit. Assamese Women in Indian Independence Movement: With a Special Emphasis on Kanaklata Baruah. Mittal Publications New Delhi; c2008

Raymond E., Wiest Jane S.P., Mocellin D. Thandiwe Motsisi, Winnipeg, & Manitoba (1994). The Needs of Women in Disaster and Emergencies. United Nations Development Programme.

Solomon Z. Wuyeba, Vincent C. Dung, Anin H. Buhari, Dalong H. Madaki, Baminda A. Bipus (2014). Women Participation in Environmental Protection & Management: Lesson from Plateau State, Nigeria. American Journal of Environmental Protection. 2(2): 32-36.

Suchinmayee, Rachna (2008). Gender, Human Rights and Environment. Atlantic Publishers & Distributors Pvt. Ltd.

16

Building Sustainable Ecosystems Fostering Collaboration and Knowledge Sharing

Gaurav Kumar, Yash Pateriya, Arvind Kumar and Gaurav Kumar

Department of Extension Education, Acharya Narendra Deva University of Agriculture and Technology, Kumarganj, Ayodhya, Uttar Pradesh, India

Abstract

This book chapter delves into the imperative of building sustainable ecosystems through the lens of fostering collaboration and knowledge sharing, with a particular emphasis on empowering rural women in the context of India. It critically examines the challenges faced by rural women in accessing information, resources, and opportunities for active participation in environmental decision-making and management. The narrative explores the transformative potential of emerging technologies in bridging the digital divide, enabling rural women to become instrumental agents of change in sustainable ecosystem development. The chapter concludes with a call to action, advocating for a collective effort involving diverse stakeholders to create an enabling environment through targeted investments in digital infrastructure, promotion of digital literacy, and culturally sensitive technological solutions. By fostering collaboration and knowledge sharing, the chapter envisions a future where rural women actively shape sustainable ecosystems, ensuring their long-term resilience.

Keywords: *Sustainable Ecosystems, Collaboration, Knowledge Sharing, Rural Women Empowerment, Emerging Technologies etc.*

16.1 Introduction

16.1.1 Definition of Sustainable Ecosystems

A sustainable ecosystem is a complex network of living organisms and their non-living environment that can maintain itself and its processes indefinitely. This includes the interaction of plant and animal life, water, air, soil, and other physical and chemical components. These interactions ensure a delicate balance that allows the ecosystem to thrive and provide essential resources for all living beings, including humans.

(*Source:*https://media.nationalgeographic.org/assets/reference/assets/ecosystem-1.pdf)

16.1.2 Importance of Collaboration and Knowledge Sharing

Collaboration and knowledge sharing are crucial cornerstones for building and maintaining sustainable ecosystems. By working together, diverse stakeholders with various expertise can develop and implement solutions that address complex environmental challenges. This collaborative approach fosters the exchange of ideas, experiences, and best practices, leading to more effective resource management, conservation efforts, and innovative solutions for environmental sustainability. (*Source:* https://www.mdpi.com/2071-1050/13/22/12788)

16.1.3 Purpose and Scope of the Chapter

This chapter, titled "Building Sustainable Ecosystems: Fostering Collaboration and Knowledge Sharing," explores the critical role of collaboration and knowledge sharing in empowering rural women in India to contribute to and benefit from sustainable ecosystems through emerging technologies. The chapter examines how digital tools and platforms can bridge the digital divide, enabling rural women to participate in knowledge sharing networks, access relevant information and resources & collaborate effectively on sustainable ecosystem management initiatives. The scope encompasses the specific challenges and opportunities rural women face in the context of sustainable ecosystems and the potential of emerging technologies to bridge these gaps.

16.2 Understanding Sustainable Ecosystems

16.2.1 Overview of Sustainable Ecosystems

A sustainable ecosystem is a complex network of living organisms (biotic) and their non-living environment (abiotic) that interacts and co-evolves to maintain a dynamic equilibrium over long periods ([Folke *et al.*, 2016]). This intricate web of life encompasses diverse species, from microscopic bacteria to towering trees, all interconnected through intricate food webs and nutrient cycles. The abiotic components, including air, water, soil, and minerals, provide the necessary foundation for life to thrive.

16.2.2 Ecological and Social Components

16. 2.2.1 Ecological Components

i. **Biodiversity:** The variety of life forms at different levels - genetic, species, and ecosystem is vital for ecosystem resilience and functionality

(Loreau *et al.*, 2001). Diverse species ensure healthy food webs, efficient nutrient cycling, and natural pest control.

ii. **Ecosystem Services:** These are the benefits humans derive from healthy ecosystems, including clean air and water, food production, climate regulation, and flood control (Costanza *et al.*, 2014). Maintaining the ecological integrity of these systems is crucial for human well-being.

16.2.2.2 Social Components

i. **Human Populations:** Humans are integral parts of ecosystems, relying on them for resources and impacting their functioning. Sustainable interactions with ecosystems require responsible management practices and conservation efforts.

ii. **Social Systems:** Cultural values, governance structures, and economic activities significantly influence how humans interact with and manage ecosystems. Building sustainable ecosystems requires collaborative efforts across diverse social groups.

16.2.3 Role of Collaboration in Ecosystem Sustainability

i. **Shared Knowledge and Expertise:** Combining knowledge from various disciplines allows for a holistic understanding of complex environmental challenges and the development of effective solutions.

ii. **Joint Decision-Making:** Collaborative decision-making that includes the voices of all stakeholders fosters more inclusive and equitable solutions for ecosystem management.

iii. **Increased Capacity and Resources:** Collaboration allows for leveraging the collective resources and capabilities of different groups to address large-scale environmental challenges.

iv. **Social Learning and Capacity Building:** Continuous knowledge exchange and mutual learning within collaborative networks empower communities and strengthen their capacity to manage their local ecosystems effectively.

16.2.4 The Significance of Collaboration: Benefits of Collaborative Approaches

Collaboration, particularly among diverse stakeholders with varying expertise and perspectives, offers a multitude of benefits when aiming to build and maintain sustainable ecosystems. These benefits can be categorised into environmental and social and economic domains.

16.3 Environmental Impact

i. **Enhanced decision-making:** Collaborative efforts bring together diverse knowledge and perspectives, leading to a more comprehensive understanding of complex environmental challenges. This facilitates informed decision-making, considering the ecological, social, and economic factors surrounding resource management and conservation efforts (Chaffin *et al.*, 2016).

ii. **Improved resource management:** Collaboration fosters collective action and shared responsibility for resource management. This can lead to more efficient and sustainable practices, such as water conservation through community-managed irrigation systems or collaborative efforts to prevent deforestation (Agrawal, 2001).

iii. **Increased innovation and knowledge sharing:** Collaboration creates platforms for sharing knowledge, best practices, and innovative solutions among stakeholders. This fosters the development and implementation of effective strategies for addressing environmental challenges, such as utilizing local knowledge for ecosystem restoration projects (Berkes, 2008).

16.3.1 Social and Economic Benefits

i. **Empowerment and capacity building:** Collaboration empowers individuals and communities by providing opportunities for skill development, leadership training, and participation in decision-making processes. This is particularly significant for marginalized groups, such as rural women, who can gain a voice in shaping sustainable practices in their communities (Agarwal, 2000).

ii. **Enhanced livelihoods and economic opportunities:** Collaborative approaches can unlock economic opportunities for individuals and communities by promoting sustainable resource management practices that support eco-tourism, sustainable agriculture, and other green enterprises (Jones *et al.*, 2018).

iii. **Social cohesion and conflict resolution:** Collaboration fosters dialogue and social cohesion, facilitating conflict resolution and fostering a sense of shared responsibility for environmental stewardship. This can be crucial in addressing complex environmental challenges that often involve competing interests and diverse perspectives (Carlsson & Berkes, 2005).

16.4.4 Knowledge Sharing in Ecosystem Building

16.4.1 Role of Knowledge in Sustainable Practices

Building and maintaining sustainable ecosystems require a deep understanding of the complex interactions within them. This knowledge empowers individuals and communities to:

i. **Develop and implement effective resource management strategies:** Knowledge about soil health, water resources, and biodiversity is crucial for promoting sustainable agricultural practices, water conservation, and ecosystem restoration.

ii. **Adapt to and mitigate the impacts of climate change:** Knowledge about climate change drivers and impacts empowers communities to adopt sustainable practices that reduce their carbon footprint and build resilience to extreme weather events. (Source: https://www.ipcc.ch/sr15/chapter/chapter-3/)

iii. **Foster innovation and collaboration:** Sharing knowledge about successful sustainable practices across communities can inspire new solutions and encourage collaborative action towards shared goals.

16.4.2 Challenges and Solutions in Knowledge Sharing

Despite its critical role, knowledge sharing in the context of ecosystem building faces several challenges, particularly in rural areas:

i. **Limited access to information and technology:** Rural communities often have limited access to reliable internet connectivity, digital devices, and relevant information resources, hindering knowledge dissemination and acquisition.

ii. **Language barriers:** Diverse languages and dialects within communities can create communication gaps and hinder knowledge exchange.

iii. **Power imbalances and gender inequality:** Traditional structures and gender inequalities can limit the participation of women and marginalized groups in knowledge-sharing processes.

iv. To address these challenges and promote effective knowledge sharing, solutions include:

v. **Developing localized, accessible knowledge resources:** Creating culturally appropriate and language-specific resources in formats accessible to diverse communities, including offline materials and low-bandwidth applications.

vi. **Promoting community-based knowledge exchange:** Facilitating peer-to-peer learning through workshops, farmer-to-farmer exchanges, and local knowledge-sharing platforms fosters inclusivity and empowers marginalised groups.

vii. **Leveraging technology for communication and collaboration:** Utilizing online platforms, mobile applications, and ICT tools adapted to local contexts can connect individuals and facilitate knowledge exchange across geographical boundaries.

16.5 Frameworks for Collaboration

Building sustainable ecosystems requires fostering collaboration and knowledge sharing among diverse stakeholders. This section explores frameworks that can empower rural women in India to participate in and contribute to these efforts through emerging technologies.

16.5.1 Building Partnerships

16.5.1.1 Interdisciplinary Collaboration

Interdisciplinary collaboration brings together individuals and groups from various backgrounds and expertise to address complex challenges. In the context of sustainable ecosystems, this can involve:

i. **Scientists, researchers, and engineers collaborating with rural communities** to understand local environmental challenges, develop appropriate solutions and ensure their cultural sensitivity and effectiveness.

   (*Source:*https://www.sciencedirect.com/science/article/pii/S0959378009000405)

ii. **Rural women collaborating with each other** to share traditional knowledge, experiences, and best practices in sustainable resource management.

   (Source:https://www.iucn.org/story/202212/indigenous-womens-insights-stewardingearth)

iii. **Digital technology experts collaborating with these stakeholders** to develop and implement user-friendly platforms and tools for knowledge sharing and communication.

16.5.1.2 Public-Private Partnerships (PPPs)

PPPs can leverage the strengths of both public and private sectors to achieve sustainable development goals. This can entail:

i. **Collaboration between government agencies and private companies** to invest in and implement digital infrastructure that improves rural connectivity and access to information and resources essential for sustainable ecosystem management.

ii. **Private companies partnering with NGOs and research institutions** to develop and deliver training programs and capacity-building initiatives that equip rural women with the necessary digital skills to participate effectively in knowledge-sharing networks and collaborative projects. (*Source:* https://www.un.org/en/desa/public-private-partnerships-and-2030-agenda-sustainable-development-fit-purpose)

16.5.2 Community Engagement

Meaningful community engagement is essential for ensuring the success of collaborative efforts. This involves:

i. **Facilitating dialogues and workshops** where rural women can voice their concerns, share their knowledge, and actively participate in decision-making processes related to sustainable ecosystem management.

ii. **Utilizing community radio, traditional communication channels, and culturally-appropriate digital platforms** to ensure inclusive information dissemination and knowledge sharing within communities. (*Source:* https://www.fao.org/forestry/participatory/en/)

16.5.3 Government and NGO Involvement

Governments and NGOs play a crucial role in enabling and supporting collaborative frameworks by:

i. **Developing policies and regulatory frameworks** that incentivise and facilitate collaboration between different stakeholders, including rural women.

ii. **Providing financial and technical assistance** to support the development and implementation of collaborative projects, capacity-building initiatives, and digital infrastructure development in rural areas.

iii. **Collaborating with research institutions and private companies** to develop innovative tools and technologies that are accessible and user-friendly for rural women to participate in knowledge sharing and collaborative projects. (Source: https://www.un.org/en/desa/public-private-partnerships-and-2030-agenda-sustainable-development-fit-purpose).

iv. By fostering collaboration through these various frameworks, rural women in India can be empowered to actively contribute to building and maintaining sustainable ecosystems while benefiting from the opportunities presented by emerging technologies.

16.6 Barriers Faced by Rural Women in India Strategies to Overcome Barriers

Table 16.1: Shows Barriers Faced by Rural Women in India Strategies to Overcome Barriers

Barriers Faced by Rural Women in India	Strategies to Overcome
Digital Divide: Limited access to ICTs, internet connectivity, and digital literacy	• Digital infrastructure development (expanding internet connectivity, establishing community access points, affordable data plans)
Socio-cultural Barriers: Gender roles, social norms, cultural biases	• Digital literacy training programs (culturally appropriate, local languages) • Gender-inclusive design of technology platforms and training programs Promoting community-based initiative
Language and Accessibility: Knowledge resources and platforms not available in local languages	• Localization of content and platforms into local languages • Developing offline and low-bandwidth resources
Lack of Awareness and Training: Limited awareness of available technologies, platforms, and opportunities	• Awareness campaigns and outreach programs • Capacity-building initiatives focused on digital skills and sustainable practices

16.7 Promoting Collaboration in Different Sectors

Building sustainable ecosystems requires collaboration across various sectors, each with unique challenges and opportunities. Here, we explore how fostering collaboration through emerging technologies can empower rural women in India to contribute to and benefit from sustainable practices in three key areas:

16.7.1 Sustainable Agriculture

i. **Challenge:** Rural women often lack access to information and resources regarding sustainable agricultural practices, hindering productivity and environmental sustainability.

ii. **Opportunity:** Digital platforms can provide access to educational resources, expert advice, and market information on sustainable farming methods, empowering women to make informed decisions. Examples include mobile apps like "Krishi Vigyan Kendra (KVK)" and "M-Kisan" offering localised information and agricultural advisories.

iii. **Collaboration:** Partnerships between government agencies, NGOs, and private companies can develop digital platforms specifically tailored to the needs of rural women farmers, addressing language barriers and providing training on using these technologies. (Source: https://www.mdpi.com/2077-0472/12/2/297)

16.7.2 Urban Planning

i. **Challenge:** Rapid urbanisation can have a detrimental impact on surrounding ecosystems, often disproportionately affecting rural communities.

ii. **Opportunity:** Engaging rural women in urban planning through digital platforms like participatory mapping tools can ensure their voices are heard, and their knowledge of local ecosystems is integrated into planning decisions. This can lead to more sustainable urban development that minimises negative impacts on rural livelihoods and ecosystems.

iii. **Collaboration:** Initiatives by urban development authorities in partnership with local women's groups and NGOs can facilitate training on using participatory mapping tools and ensure their meaningful participation in the planning process. (Source: https://www.undp.org/sdg-accelerator/tools)

16.7.3 Biodiversity Conservation

i. **Challenge:** Rural women often possess valuable knowledge of local biodiversity and traditional conservation practices, but lack opportunities to share and leverage this knowledge for broader conservation efforts.

ii. **Opportunity:** Online platforms and digital storytelling tools can enable women to document and share their knowledge about local flora and fauna, contributing to community-based conservation initiatives. Citizen science apps, such as "iNaturalist," can further empower women to collect and contribute valuable data on biodiversity.

iii. **Collaboration:** Partnerships between research institutions, conservation NGOs, and local women's groups can facilitate knowledge exchange and capacity building, enabling women to use technology effectively for biodiversity conservation and become active participants in scientific research. (Source: https://www.mdpi.com/2071-1050/15/14/11104)

16.8 Building a Culture of Knowledge Sharing

Cultivating a culture of knowledge sharing is vital for empowering rural women to play a significant role in building sustainable ecosystems. This section explores three key avenues that can foster this crucial exchange of knowledge and empower women:

16.8.1 Educational Initiatives

i. **Formal education:** Integrating themes of sustainable ecosystems and the use of digital tools for knowledge sharing into the formal education curriculum can equip young girls and women with the necessary knowledge and skills to participate meaningfully in environmental efforts. This could involve incorporating modules on sustainable agriculture, environmental conservation, and digital literacy into school curricula at various levels (Lang & Wiek, 2012).

ii. **Non-formal and informal education:** Providing targeted non-formal and informal educational programs focused on sustainable practices and digital literacy can equip rural women with the knowledge and skills needed to contribute to sustainable ecosystem management. This could include workshops, training sessions, and community outreach programs conducted in local languages and adapted to specific cultural contexts (Bawane *et al.*, 2020).

16.8.2 Corporate Practices

i. **Knowledge sharing platforms:** Corporations operating in rural areas can contribute by developing and disseminating knowledge sharing platforms accessible to rural women. These platforms can host information on sustainable practices, local success stories, and best practices in resource management. Additionally, they can facilitate peer-to-peer learning and communication among women (World Bank, 2017).

ii. **Partnerships with NGOs and government agencies:** Corporations can collaborate with NGOs and government agencies to develop and implement joint initiatives aimed at promoting knowledge sharing and capacity building among rural women. This collaborative approach can leverage the expertise and resources of different stakeholders to achieve greater impact (Prahalad, 2005).

16.8.3. Grassroots Movements

i. **Women-led knowledge networks:** Supporting the formation and growth of women-led knowledge networks on sustainable practices can foster horizontal knowledge sharing and empower women to learn from each other's experiences and expertise. These networks can be facilitated through digital platforms, community meetings, and knowledge exchange programs (Agarwal, 2000).

ii. **Local knowledge repositories:** Encouraging the creation and maintenance of local knowledge repositories documenting traditional ecological knowledge and best practices can preserve valuable knowledge and promote its dissemination among younger generations. This can involve collaborating with local communities to document this knowledge in accessible formats (Tengo *et al.,* 2014).

16.9 Future Outlook

16.9.1 Emerging Trends in Sustainable Ecosystem Development

The future of sustainable ecosystem development holds promising advancements fueled by technological innovation and a growing emphasis on collaborative solutions. Some key emerging trends include:

i. **Digitalization:** The integration of digital tools like artificial intelligence, big data analytics, and the Internet of Things (IoT) will play a crucial role in monitoring, analysing, and managing ecosystems more effectively. This can lead to improved resource use, early detection of environmental threats, and targeted conservation efforts. *(Source: https://www.undp.org/arab-states/digital-sustainable-development)*

ii. **Nature-based solutions:** There is a growing focus on utilising natural processes to address environmental challenges. This includes strategies like ecosystem restoration, reforestation, and green infrastructure development, all of which play a vital role in maintaining healthy ecosystems\ *(Source: https://portals.iucn.org/library/sites/library/files/documents/2020-020-En.pdf)*

iii. **Circular economy:** The shift towards circular economies, where waste is minimised and resources are reused and recycled, will be crucial for achieving long-term sustainability. Rural communities can play a vital role in developing and implementing circular economy models at the local level. *(Source: https://www.ellenmacarthurfoundation.org/topics/circular-economintroduction/overview)*

16.9.2 Opportunities for Future Collaboration

Collaboration will continue to be a critical driver of progress in building sustainable ecosystems. Here are some promising avenues for future collaboration:

i. **Strengthening multi-stakeholder partnerships:** Fostering collaboration between government agencies, NGOs, research institutions, private sector entities, and local communities is essential for developing and implementing effective solutions. This necessitates inclusive decision-making processes that incorporate diverse perspectives and local knowledge. *(Source: https://www.undp.org/sites/g/files/zskgke326/files/2022-11/UNDP-UNHCR-Global-Collaboration-Framework-for-Inclusion-and-Solutions-2023-2025.pdf)*

ii. **Promoting knowledge co-creation and exchange:** Encouraging knowledge exchange platforms that connect rural women with scientists, experts, and other stakeholders can foster knowledge cocreation and build collective capacities to address local environmental challenges effectively. *(Source:https://www.sciencedirect.com/science/ article/pii/ S2210422419302655)*

iii. **Leveraging technology for inclusive collaboration:** Digital tools and platforms can bridge the geographical divide and facilitate communication and collaboration between diverse stakeholders, including rural women, even in remote locations. This can empower them to share their knowledge, experiences, and perspectives, fostering a more inclusive approach to sustainable ecosystem management..

(Source: https://en.unesco.org/sustainabledevelopmentgoals)

16.10 Conclusion

In conclusion, this chapter underscores the pivotal role of collaboration and knowledge sharing in fostering sustainable ecosystems, with a particular focus on empowering rural women in India. It highlights the challenges faced by these women in accessing information and resources for meaningful participation in environmental decision-making. The discussion also explores the transformative potential of emerging technologies in overcoming these challenges, bridging the digital divide, and enabling rural women to proactively contribute to building sustainable ecosystems. The call to action emphasises the necessity of a collective effort involving policymakers, researchers, civil society organisations, and local communities to create an enabling environment. This involves targeted investments in digital infrastructure, the promotion of digital literacy, and the development of culturally appropriate technological solutions. The chapter concludes with reflections on empowering rural women through digital technologies, emphasising the importance of collaboration and knowledge sharing in shaping a future where they are not merely beneficiaries but active participants, ensuring their long-term resilience and contributing to sustainable ecosystem development.

References

Agarwal, A. (2001). Common property regimes, environmental protection, and participation: The case for community-controlled water management in Madhya Pradesh, India. Economic andPoliticalWeekly,36(48):4601-4608https://link.springer.com/article/10.1007/BF00324686

Agarwal, B. (2000). Gender, poverty and environmental degradation. Studies in Human Ecology, 8(1): 167-178. https://www.researchgate.net/publication/288906123_Rural_Women_Poverty_and_Natural_Resources_Sustenance_Sustainability_and_Struggle_for_Change

Agarwal, B. (2000). Omkara: Gender equity, empowerment, and sustainable development. Zed Books Ltd.

Agarwal, B. (2000). The gender and environment debate: Lessons from India. Feminist Studies, 26(1): 119-168.

Agarwal, S. (2023, January 11). Bridging the digital divide in rural India: Challenges and opportunities.https://www.researchgate.net/publication/352442075_Bridging_digital_divide_in_India_Way_forward_challenges

Bawane, M., Bawane, S., & Sharma, R. (2020). Impact of non-formal education on the sustainability of women self-help groups in rural India. International Journal of Scientific & Engineering Research, 11(5), 1580-1586.

Berkes, F. (2008). Sacred ecology (2nd ed.). Routledge.

Carlsson, L., & Berkes, F. (2005). Co-management: Collective action and learning in complex social-ecological systems. Routledge.

Chaffin, B. C., Gosselink, T. E., Hicks, M. D., & Myer, J. H. (2016). A framework for assessing the social-ecological effects of conservation decision-making. Environmental Management,57(5),778-794.https://wilsonconservationecology.com/our-research/research-themes/conservation-decision-making-in-social-ecological-systems/

Costanza, R., de Groot, R., van der Ploeg, S., et al. (2014). Changes in the global value of ecosystem services. Global Environmental Change, 26(1), 152-158. https://www.sciencedirect.com/ science/article/abs/pii/S0048969721062008

Das, M., & Joshi, P. K. (2017). Role of women in sustainable development: A case study of India.Journal of Management and Society, 3(2): 119-131.

Ellen MacArthur Foundation. (https://www.ellenmacarthurfoundation.org/topics/circular-economy-introduction/overview) What is a Circular Economy?

Folke, C., Biggs, R., Norstrom, A., et al. (2016). Social-ecological resilience and planetary boundaries. Nature Sustainability, 1(7): 1-9. https://www.nature.com/articles/d41586-023-01749-9

Gupta, S., & Gupta, S. (2020). Role of digital divide in rural India. https://www.researchgate.net/publication/254120418_The_Rural-Urban_Digital_Divide

https://www.mdpi.com/2071-1050/15/14/11104

https://www.sciencedirect.com/science/article/pii/S0743016723000852

https://www.undp.org/sdg-accelerator/tools

International Union for the Conservation of Nature (IUCN). (https://portals.iucn.org/library/sites/library/files/documents/2020-020-En.pdf) Nature-based Solutions.

Jones, P., van der Ploeg, J., & Pimbert, M. (2018). Environmental governance, social innovation and collective action. Routledge.

Labiotech.eu. (2023, October 26). Open ecosystems promote knowledge sharing and scientific innovation.https://www.labiotech.eu/partner/innovation-ecosystems-encourage-knowledge-sharing-scientific-innovation/

Lang, D. J., & Wiek, A. (2012). Bridging the gap between science and society: The potential of knowledge co-production for sustainable development. Environmental Science & Policy, 21(4): 449-459.

Loreau, M., Naeem, S., Inchausti, P., et al. (2001). Biodiversity and ecosystem functioning: Current knowledge and future challenges. Science, 294(5543): 1120-1126. https://www.science.org/doi/10.1126/science.1217245

McEvoy, Julia. (2014). Social learning for transitions. The Journal of Environmental Planning and Management, 58(7): 1089-1112.

National Geographic Society. (n.d.). Sustainable ecosystem. https://media.nationalgeographic.org/assets/reference/assets/ecosystem-1.pdf

Prahalad, C. K. (2005). The fortune at the bottom of the pyramid: Eradicating poverty through profits. Pearson Education.

Tengo, M., Brondizio, E., Elmqvist, T., Malmstrom, P., & Nystrom, M. (2014). Integrating societal and ecological knowledge for resilience in social-ecological systems. Ambio, 43(5): 572-580.

United Nations Development Programme (UNDP). (2020). Gender and environment. https://www.undp.org/pacific/gender

United Nations Development Programme (UNDP). (https://www.undp.org/arab-states/digital-sustainable-development) Digitalization for Sustainable Development.

United Nations Development Programme (UNDP). (https://www.undp.org/sites/g/files/zskgke326/files/2022-11/UNDP-UNHCR-Global-Collaboration-Framework-for-Inclusion-and-Solutions-2023-2025.pdf) A Framework for Collaboration and Innovation in Sustainable Development Challenges.

UN Women. (2023, May 31). Internet access for rural women in India. https://www.unwomen.org/en/news-stories/op-ed/2022/09/expanding-broadband-safely-and-inclusively-to-reach-digital-gender-equality

World Bank (2017). World Development Report 2017: Governance and the Law. The World Bank.

World Economic Forum. (2020, December 15). A gender gap in the digital divide. https://www.weforum.org/agenda/2022/08/womens-education-digital-divide/

17

Data Privacy and Security A Critical Lens on Gender and Technology

Ishita Mishra[1], Gaurav Beri[1], Disha Agarwal[1], Amrit Warshini[2] and Aman Verma[2]

[1]*Department of Agricultural Communication, G.B Pant University of Agriculture and Technology, Pantnagar, Uttarakhand, India*

[2]*Department of Extension Education, Acharya Narendra Deva University of Agriculture and Technology Kumarganj, Ayodhya, Uttar Pradesh, India*

Abstract

Data privacy and security are paramount in our technology-driven world, yet the conversation often overlooks the gender dimension. This chapter argues for a critical lens that examines how data practices intersect with gender identity. It explores how existing gender inequalities are mirrored and amplified online, with patriarchal structures impacting access, control, and agency in the digital sphere. The chapter delves into the gendered implications of emerging technologies like artificial intelligence (AI) and machine learning, investigating how algorithms can perpetuate or challenge gender biases. By integrating gender perspectives into data privacy and security discussions, the chapter emphasises the need for gender-inclusive approaches in technological development and policy making. This fosters a vision for more equitable digital ecosystems where everyone, regardless of gender, can control their data and thrive in the digital age.

Keywords: *Data privacy, Security, Gender, Gender inequality, Digital sphere, Artificial intelligence, Machine learning, Algorithmic bias, Gender-inclusive policy.*

17.1 Introduction

In an era where digital technologies permeate every facet of our lives, the discourse surrounding data privacy and security has never been more pertinent. However, within this discourse lies a critical blind spot the intersectionality of gender and technology. As we navigate the complexities of safeguarding personal data and ensuring online security, it is important to recognise and address the differential impacts experienced by individuals based on their gender identities. The combination of gender dynamics with data privacy and security unveils a multifaceted landscape characterised by unique challenges and opportunities. This intersectionality necessitates an examination that goes beyond conventional approaches, requiring a critical lens that acknowledges the diverse experiences and power dynamics at play.

The present chapter highlights the relationship between data privacy, security, and gender, offering insights into how prevailing inequalities are replicated and exacerbated within digital realms. The chapter explores how patriarchal structures influence access, control, and agency in the online sphere and interrogates the implications of emerging technologies such as artificial intelligence (AI) and machine learning through a gendered lens, uncovering the ways in which algorithmic decisions perpetuate or challenge existing gender norms and biases. By doing so, the aim is to elucidate the importance of integrating gender perspectives into discussions surrounding data privacy and security and highlight the necessity for gender-inclusive approaches in technological development and policy-making. By centering gender within the discourse on data privacy and security, we aspire to foster more equitable and inclusive digital ecosystems, where individuals of all genders can assert control over their personal data and participate fully in the digital age.

17.2 Data Privacy

Data privacy refers to individuals' ability to determine the circumstances under which their personal information is shared with others, including what information is shared, how it is shared, and to what extent. This personal information encompasses details like one's name, location, contact details, or online and offline behaviours. Much like controlling access to a private conversation, many internet users seek to manage or restrict the collection of specific types of personal data. With the rise in internet usage, the significance of data privacy has also grown. Websites, applications, and social media platforms often require personal data from users to deliver services effectively. However, some platforms may go beyond users' expectations in their data collection and utilisation, resulting in decreased privacy. Additionally, inadequate safeguards around collected data can lead to data breaches, further compromising user privacy. In numerous legal jurisdictions, privacy is recognised as a fundamental human right, and laws governing data protection aim to uphold this right. Data privacy is crucial because individuals must trust that their personal information will be handled responsibly in order to engage online. Organisations implement data protection measures to assure their customers and users that they can be relied upon to safeguard their personal data. If personal data is not kept private or individuals lack control over its use, various forms of misuse can occur:

- Criminals might exploit personal data for fraudulent activities or harassment.
- Entities may unlawfully sell personal data to advertisers or third parties, leading to unwanted marketing or advertising for users.
- Surveillance and tracking of individuals' activities could impede their freedom of expression, particularly in repressive governmental contexts.
- These outcomes can be detrimental for individuals, while for businesses, they can result in irreparable damage to their reputation, as well as legal penalties such as fines and sanctions.

Beyond the practical consequences of privacy breaches, many individuals and nations argue that privacy holds intrinsic value: it is considered a fundamental human right essential for a free society, akin to the right to free speech.

17.3 Understanding the Changing Context of Digital Security through a Gender Lens

In recent years, the digital landscape has undergone rapid transformation, shaping various aspects of our lives from communication to commerce, education to entertainment. However, as digital technologies become increasingly pervasive, the issue of digital security has garnered significant attention. While discussions around digital security often focus on technical aspects such as encryption and cybersecurity measures, it is essential to adopt a gender lens to understand the nuanced ways in which digital security intersects with gender dynamics. This essay explores the changing context of digital security through a gender perspective, highlighting the unique challenges and opportunities faced by women in the digital realm. Digital technologies have the potential to empower individuals and communities, providing access to information, services, and opportunities like never before. However, gender disparities in access to and control over digital technologies persist, limiting women's ability to fully benefit from the digital revolution. According to the World Wide Web Foundation, globally, women are 23% less likely than men to use mobile internet, with this gap widening in regions such as sub-Saharan Africa and South Asia. Additionally, women are underrepresented in STEM fields, further exacerbating inequalities in digital literacy and skills.

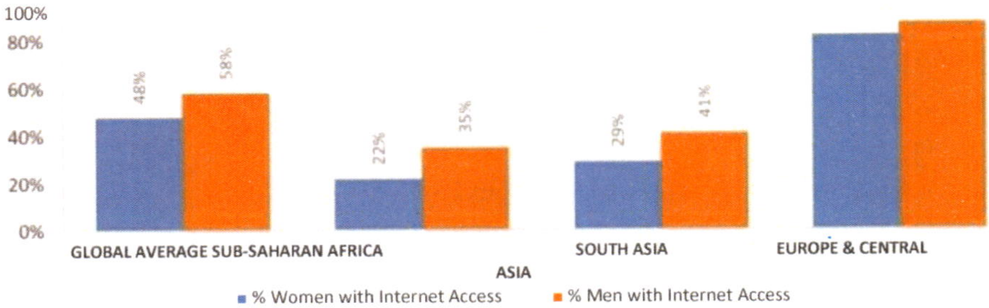

Fig. 17.1: Internet access by gender
(*Source:*https://www.gsma.com/r/wp-content/uploads/2022/06/The-Mobile-Gender-Gap-Report- 2022.pdf)

The lack of digital access and literacy among women not only hinders their economic and social opportunities but also exposes them to various digital security risks. For example, women are more likely to experience online harassment and cyberstalking, with research showing that women are disproportionately targeted with abusive and threatening messages on social media platforms. Furthermore, intimate partner violence often extends into the digital realm, with abusers using technology to monitor, control, and intimidate their victims. Moreover, the gender digital divide intersects with other forms of inequality, such as race, ethnicity, class, and sexual orientation, further exacerbating disparities in digital security. For instance, marginalised women, including women of colour, low-income women, and LGBTQ+ individuals, are more vulnerable to online abuse and discrimination. Intersectional approaches are thus crucial in addressing the multifaceted challenges faced by different groups of women in the digital space.

In addition to the threats posed by digital inequality, the rapid pace of technological innovation presents new challenges for digital security. Emerging technologies such as artificial intelligence (AI), biometrics, and the Internet of Things (IoT) raise concerns about privacy, data protection, and algorithmic bias, which can disproportionately affect women and marginalised groups. For example, biased algorithms used in recruitment processes may perpetuate gender stereotypes and hinder women's access to employment opportunities. However, while digital technologies present challenges, they also offer opportunities for promoting gender equality and enhancing digital security. Initiatives such as digital literacy programs, women-led tech start-ups, and online support networks can empower women to navigate the digital landscape safely and assert their rights online. Furthermore, advocating for gender-responsive policies and regulations can ensure that digital platforms and services are designed with women's needs and safety in mind.

Addressing the changing context of digital security through a gender lens requires a multi-stakeholder approach involving governments, civil society, tech companies, and the broader community. Policymakers must prioritise bridging the gender digital divide by investing in infrastructure, education, and skills development initiatives that are inclusive and accessible to all. Tech companies have a responsibility to design products and services that prioritise user privacy and security, with a particular focus on addressing the needs of women and marginalised groups. Civil society organisations play a crucial role in raising awareness about digital security risks and providing support and resources to victims of online harassment and abuse. Community-driven initiatives such as digital literacy workshops, cyber self-defence training, and online safety campaigns can empower women to assert their digital rights and protect themselves in the digital space.

Understanding the changing context of digital security through a gender lens is essential for addressing the complex challenges and opportunities posed by the digital revolution. By recognising the intersecting inequalities that shape women's experiences online and adopting inclusive and participatory approaches, we can work towards creating a digital world that is safe, equitable, and empowering for all.

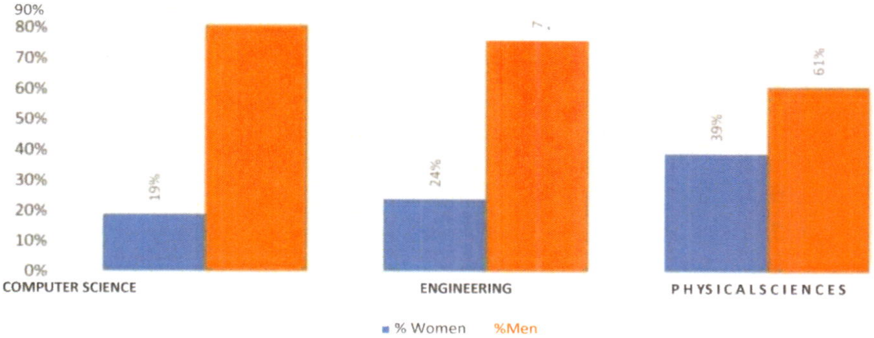

Fig. 17.2: Representation in STEM fields

(*Source:* https://en.wikipedia.org/wiki/Women_in_STEM_fields)

17.3.1 Gender Perspective to Data Security

In exploring the nexus of gender, technology, and data privacy, it becomes evident that existing inequalities are replicated and new social injustices emerge due to the decisions made by algorithms within AI and machine learning systems. Adopting a feminist perspective facilitates a comprehensive examination of the spectrum of issues and opportunities stemming from these technologies within the context of gender inequality. This feminist approach is instrumental in addressing the fundamental question of why gender is a crucial factor in discussions surrounding privacy and data protection.

Through the lens of feminism, it becomes apparent that women and gender-diverse individuals encounter "uneasy access" within the online sphere, reflecting patriarchal norms that seek to control their visibility. Emerging issues such as data veillance by private entities, businesses, and society at large further exacerbate these inequalities. Additionally, biases and discrimination embedded within algorithms contribute to violations of human dignity, as individuals experience a loss of agency and control in their interactions with technology. These power dynamics not only reflect digital inequalities but are also shaped by offline realities. By examining data privacy and security through a feminist lens, we gain deeper insights into the ways in which gender dynamics intersect with technological advancements. This critical perspective underscores the urgency of addressing these issues to foster more equitable and inclusive digital environments.

17.3.2 Understanding Gender Dynamics in the Digital Age

In the digital age, gender dynamics have undergone a profound transformation, influenced by technological advancements, social media, and evolving cultural norms. This shift has brought both challenges and opportunities, shaping how individuals perceive, express, and experience gender identity in virtual and real-life spaces.

i. **The Impact of Social Media:** Social media platforms have become powerful arenas for shaping gender norms and expectations. They offer spaces for individuals to explore and express their gender identities freely, but they also perpetuate stereotypes and create pressure to conform to traditional gender roles. The curated nature of social media profiles can exacerbate feelings of inadequacy and insecurity, particularly among young people striving for unattainable beauty standards or conforming to rigid gender norms.

ii. **Cyberbullying and Harassment:** Unfortunately, the digital age has also amplified instances of cyberbullying and harassment, often targeting individuals based on their gender identity. Women and gender minorities, in particular, face disproportionate levels of online abuse, ranging from sexist remarks to threats of violence. This pervasive harassment not only undermines individuals' sense of safety and autonomy but also perpetuates a culture of fear and silence.

(*Source:* https://www.datasociety.net)

iii. **Online Activism and Empowerment:** Despite these challenges, the digital age has provided a platform for marginalised voices to amplify their stories and advocate for gender equality. Online activism campaigns, such as #MeToo and #BlackLivesMatter, have sparked global conversations about gender-based violence and systemic discrimination. Social media has enabled individuals to connect with like-minded communities, organise grassroots movements, and challenge oppressive structures, fostering a sense of solidarity and empowerment.

iv. **Intersectionality and Inclusivity: Intersectionality**—a framework that recognises the interconnected nature of social identities, such as gender, race, sexuality, and class—plays a crucial role in understanding gender dynamics in the digital age. Intersectional perspectives highlight the diverse experiences of individuals and the intersecting forms of discrimination they may face. It calls for more inclusive representations and policies that consider the complexities of identity and privilege within online spaces.

v. **The Role of Algorithms and Bias:** However, the digital landscape is not immune to bias and discrimination. Algorithms used by social media platforms and search engines often reflect and reinforce existing inequalities, perpetuating gender-based stereotypes and discrimination. Biased algorithms can impact everything from job recommendations to criminal justice outcomes, exacerbating systemic inequalities and limiting opportunities for marginalised groups.

vi. **Digital Privacy and Consent:** Issues of digital privacy and consent are also central to understanding gender dynamics online. The proliferation of dating apps, for example, raises concerns about consent and the commodification of intimacy. Moreover, the misuse of personal data can have severe consequences, such as revenge porn and online stalking, disproportionately affecting women and gender minorities. Ensuring robust privacy protections and consent mechanisms is essential for creating safe and equitable digital spaces.

vii. **Education and Digital Literacy:** Addressing gender dynamics in the digital age requires comprehensive education and digital literacy initiatives. By equipping individuals with the skills to critically evaluate online content, challenge harmful stereotypes, and navigate privacy settings, we can empower them to assert control over their digital identities and experiences. Education also plays a vital role in fostering empathy, promoting respectful online behaviour, and combating cyberbullying and harassment.

Understanding gender dynamics in the digital age requires a nuanced understanding of the intersectional nature of identity, the impact of technology on social norms, and the importance of education and advocacy. While the digital landscape presents both challenges and opportunities, it is essential to strive for more inclusive and equitable online spaces where individuals of all genders can express themselves freely without fear of discrimination or harassment. By addressing systemic biases, promoting digital literacy, and advocating for meaningful change, we can create a more just and inclusive digital future.

17.4 Promoting Equality and Security in Digital Landscape

In the ever-evolving digital landscape, the intertwined issues of equality and security remain at the forefront of societal discourse. As technology continues to shape our lives in profound ways, it is imperative that we address the inherent inequalities and vulnerabilities that persist, particularly concerning gender dynamics. With data privacy breaches and cybersecurity threats on the rise, fostering an environment of equality and security is not only a moral imperative but also essential for the advancement of society as a whole. One of the key challenges in promoting equality and security in the digital landscape is the recognition and understanding of how gender intersects with technology. Historically, technology has been designed and developed with a predominantly male perspective, leading to biases and discriminatory practices that disproportionately affect women and non-binary individuals. From biased algorithms to inadequate representation in the tech industry, the digital divide is not merely about access to technology but also about the equitable distribution of its benefits and protections. Data privacy is a fundamental aspect of promoting equality in the digital age. However, women and marginalised communities often face heightened risks of privacy violations due to various factors, including online harassment, data exploitation, and surveillance. For example, studies have shown that women are more likely to experience cyberstalking and revenge porn, highlighting the urgent need for gender-sensitive approaches to data protection. Moreover, the intersectionality of gender with other factors such as race, sexuality, and socioeconomic status further exacerbates these vulnerabilities, underscoring the importance of holistic and inclusive solutions.

Similarly, ensuring cyber security is essential for creating a safe and equitable digital environment. Yet, women are significantly underrepresented in cyber security roles, constituting only a fraction of the workforce in this critical field. This lack of diversity not only perpetuates gender disparities but also hampers innovation and undermines the effectiveness of cyber security measures.

By fostering diversity and inclusivity in the cyber security sector, we can harness a wide range of perspectives and expertise to develop more robust and comprehensive security protocols. To promote equality and security in the digital landscape, policymakers, industry leaders, and civil society must collaborate to enact meaningful change. This includes implementing gender-aware policies and regulations that protect individuals' privacy rights and hold perpetrators of online harassment and abuse accountable. Additionally, investing in education and training programs can help bridge the gender gap in technology and cyber security, empowering women and marginalised groups to participate fully in the digital economy. Furthermore, fostering a culture of digital literacy and empowerment is crucial for enhancing individuals' awareness of their rights and responsibilities online. By providing resources and support networks, we can empower individuals to protect themselves against cyber threats and advocate for their privacy rights. Moreover, promoting ethical standards and responsible use of technology can help mitigate the negative impact of digital technologies on vulnerable populations, fostering a more inclusive and equitable digital society.

Promoting equality and security in the digital landscape requires a multi-faceted approach that addresses the complex interplay of gender dynamics, technology, and data privacy. By recognising and addressing the unique challenges faced by women and marginalised communities, we can create a more inclusive and secure digital environment for all. Through collaborative efforts across sectors, we can harness the transformative potential of technology to advance equality, empower individuals, and safeguard our collective digital future.

17.5 Policy and Practice: Advocating for Gender-Aware Approaches to Data Protection

In today's digital landscape, where data is ubiquitous and technology permeates every aspect of our lives, the need for robust data protection measures is more critical than ever. However, in advocating for these measures, it is essential to recognise and address the gender disparities and vulnerabilities that exist within current data protection policies and practices. Gender-aware approaches to data protection are essential for promoting equality and security in the digital realm. At the heart of gender-aware data protection lies the recognition that individuals' experiences of privacy and security in the digital world are shaped by their gender identities and roles. Women, non-binary individuals, and other marginalised gender groups often face unique challenges and risks concerning their digital privacy and security. These challenges can stem from various factors, including societal norms, cultural expectations, and systemic biases embedded within technology and data practices.

One key aspect of advocating for gender-aware data protection is ensuring that policies and practices consider the diverse needs and experiences of all genders. This requires a comprehensive understanding of how gender intersects with other axes of identity, such as race, ethnicity, sexuality, and socioeconomic status. By taking an intersectional approach, policymakers and practitioners can develop more inclusive and effective data protection measures that address the specific concerns of marginalised communities. Additionally, promoting gender-aware data protection involves challenging and dismantling gender stereotypes and biases embedded within technology and data practices. From algorithmic bias to the design of digital platforms, gender-based discrimination can manifest in various forms within the digital realm. By raising awareness of these issues and advocating for more inclusive design principles and ethical guidelines, we can work towards creating a digital environment that respects and protects individuals' rights regardless of gender.

Furthermore, advocating for gender-aware data protection requires active engagement with policymakers, industry stakeholders, and civil society organisations to integrate gender considerations into legislative frameworks, industry standards, and corporate practices. This includes advocating for stronger legal protections against gender-based discrimination and harassment online, as well as promoting transparency and accountability in data collection, processing, and sharing practices. Education and awareness-raising are also crucial components of promoting gender-aware data protection. By providing individuals with the knowledge and skills to navigate the digital world safely and securely, we can empower them to protect their privacy and security effectively. This includes educating users about their rights regarding data privacy, as well as raising awareness of the risks of gender-based online harassment and abuse. Finally, fostering collaboration and partnership among diverse stakeholders is essential for advancing gender-aware data protection efforts. By bringing together policymakers, technology companies, civil society organisations, and academic institutions, we can leverage collective expertise and resources to develop innovative solutions and best practices for promoting equality and security in the digital landscape. Advocating for gender-aware approaches to data protection is essential for promoting equality and security in the digital landscape. By recognising and addressing the unique challenges and vulnerabilities faced by marginalised gender groups, we can work towards creating a more inclusive and equitable digital environment for all. Through comprehensive policy measures, inclusive design principles, education, and collaboration, we can build a future where everyone can participate in the digital world safely and securely, regardless of gender.

17.6 Conclusion

In conclusion, we can say that the intersection of data privacy, security, gender, and technology unveils a complex web of challenges and opportunities. Through this critical lens, it becomes evident that existing technological infrastructures often fail to address the nuanced ways in which gender identity intersects with data privacy and security. Women, non-binary individuals, and other marginalised groups are disproportionately affected by privacy breaches, data biases, and online harassment. Furthermore, the pervasive gender digital divide exacerbates these issues, leaving many women and non-binary individuals at a disadvantage in navigating and securing their digital lives. Efforts to bridge this gap must involve not only improving access to technology but also addressing the systemic biases embedded within digital systems. Moreover, the commodification of personal data poses a significant threat to individual autonomy and privacy rights, with women often bearing the brunt of targeted advertising, algorithmic discrimination, and data exploitation. Stricter regulations and ethical guidelines are imperative to safeguarding the privacy and security of all individuals, regardless of gender. Advancements in technology also offer opportunities for empowerment and resistance. From encrypted communication tools to decentralised platforms, innovative solutions are emerging to protect users' privacy and challenge dominant narratives perpetuated by biased algorithms. However, addressing the multifaceted challenges of data privacy and security through a gender-sensitive approach requires collaboration across various sectors. Governments, tech companies, civil society organisations, and individuals must work together to promote digital literacy, enact inclusive policies, and hold accountable those who perpetuate harm through data exploitation. Ultimately, achieving a more equitable and secure digital landscape requires not only technical solutions but also a fundamental shift in societal attitudes towards gender and technology. By centring the experiences and needs of marginalised groups, we can strive towards a future where data privacy and security are not privileges but inherent rights for all.

References

https://datasociety.net/library/privacy-security-and-digital-inequality/https://press.un.org/en/2023/wom2227.doc.htm

World Economic Forum. The Global Gender Gap Report (2017). Available online at: http://www3.weforum.org/docs/WEF_GGGR_2017.pdf

Women for Women, "Three Ways We Can Help Refugees Move Forward," Our Blog. Accessed on March 25, 2018. https://www.womenforwomen.org/blog/three-ways-we-can-help refugees-move-forwar

18

Challenges and Oppurtunities Unpacking the Rural Landscape

***Smita Singh*[1*], *Aman Verma*[1], *Ramesh Chand Bunkar*[2] *Shivam Singh*[3], *Ritesh Singh*[1] and *Kapil Verma*[1]**

[1]*Department of Extension Education. Acharya Narendra Deva University of Agriculture and Technology, Kumarganj, Ayodhya, Uttar Pradesh, India*

[2]*Division of Dairy Extension, National Dairy Research Institute, Karnal, Haryana India*

[3]*Department of Agricultural Extension Education, Chandra Shekhar Azad, University of Agriculture & Technology, Kanpur, Uttar Pradesh, India*

Abstract

Rural communities have social, economic, and environmental challenges. Inequalities in healthcare and education are social issues, and market accessibility and outdated farming practices inhibit economic viability. Environmental vulnerabilities that endanger livelihoods as a whole and agriculture specifically are made worse by climate change. Nevertheless, there are many opportunities in the rural setting in addition to its challenges. Thanks to current technologies, agricultural practices and economic productivity can be improved. Entrepreneurial ventures based on local resources can drive economic progress, and education is essential for sustainable development in order to empower communities. Opportunities for revenue diversification while protecting natural and cultural heritage are made possible by the growing popularity of eco-friendly practices and agrotourism. Using renewable energy offers a twofold benefit: it helps to address economic issues and advance environmental sustainability. But the rural environment isn't just filled with difficulties—it also offers plenty of opportunity. With the use of modern technologies, agricultural practices and economic productivity can be enhanced. Entrepreneurial ventures based on local resources have the potential to promote economic growth, and education is a critical component of community empowerment for sustainable development. The rising acceptance of eco-friendly techniques and agrotourism opens up opportunities for revenue diversification while preserving natural and cultural assets. There are two advantages to using renewable energy: it promotes environmental sustainability and addresses economic challenges.

Keywords: *Agrotourism, Rural landscape, Agriculture, Healthcare.*

18.1 Introduction

Unpacking the rural landscape having various challenges due to its complex and composite nature.

The development of successful plans and policies requires an awareness of the following issues.

18.1.1 Heterogeneity of Rural Areas

Rural landscapes are diverse, exhibiting variations in geography, culture, and economic activities. Addressing the unique needs of different regions requires a modulation approach (Cloke & Milbourne, 1992). Different topographies, such as hills and plains, have an effect on farming practices. Economic heterogeneity results from the use of various farming techniques and the presence of various sectors. Socially, these locations are characterised by diverse ethnic groups with a diversity of cultures. Access to healthcare and education is restricted by inequalities in infrastructure and services, which add to the uneven development.

18.1.2 Data Accessibility and Quality

Accessing reliable and comprehensive data on rural areas can be challenging due to factors such as remote locations, limited resources, and outdated data collection methods (Hemmerling *et al.*, 2017). It is difficult to effectively assess rural needs and provide focused treatments when there is a lack of data.

18.1.3 Infrastructure Shortage

Rural areas often suffer from inadequate infrastructure, including roads, electricity, and telecommunications networks (Rasul & Sharma, 2016). A major barrier to growth in the economy and development is the deficiency of infrastructure. Insufficient funding for infrastructure related to communication, electricity, water, and transportation may reduce productivity, disrupt supply chains, and make it more difficult for people to access essential services. Resolving this deficit is essential to raising living standards, encouraging sustainable growth, and ensuring long-term economic stability.

18.1.4 Economic Diversification

Dependence on agriculture as the primary source of livelihood poses challenges in achieving economic diversification in rural areas (Chen & Huffman, 2019). Limited access to markets, Access and competition, Skilled Workforce Shortage, Dependency on Primary Industries, low productivity, Limited technology adoption, vulnerability to climate change, Policy and regulation barriers and resistance to change and many others further compound the economic challenges faced by rural communities.

18.1.5 Social Services and Healthcare Access

Rural communities often lack access to quality healthcare services, educational facilities, and other social amenities (Hossain *et al.*, 2020). Geographic remoteness, Financial barriers, Preventive care challenges, Limited Telehealth infrastructure, Transportation Barriers, Limited educational opportunities, Limited job opportunities, shortage of healthcare professionals, and inadequate infrastructure contribute to disparities in service delivery.

18.1.6 Environmental Sustainability

Rural landscapes are vulnerable to environmental degradation due to factors such as deforestation, soil erosion, and water pollution (VanderZanden *et al.*, 2019). Agricultural practices, Natural Resources depletion, Water scarcity and quality, Energy dependency on nonrenewable, Waste management, Climate change vulnerability, Limited environmental awareness, Land use changes, Limited access to environmental technology, and Policy and regulatory gaps can increase environmental challenges and threaten long-term viability.

18.1.7 Policy Implementation and Governance

Inadequate policy implementation and governance mechanisms can hinder effective resource allocation and service delivery in rural areas (Shahbaz *et al.*, 2019). Weak institutional capacities, corruption, and lack of community participation often lack the effectiveness of rural development initiatives.

18.1.8 Migration Trends

The migration of youth to urban areas in search of better opportunities results in demographic imbalances and the loss of skilled labour in rural regions (Rigg, 2006). People usually migrate from rural areas to urban areas in search of jobs or in search to earn money for a better lifestyle.

18.1.9 Policy Implementation

Implementing effective policies at the local level requires overcoming bureaucratic hurdles and ensuring that initiatives are tailored to the specific needs of each rural community (Cloke & Edwards, 1986). Key challenges associated with policy implementation in rural areas are limited infrastructure, Resources constraints, Lack of local capacity, Geography dispersion, Community diversity, Limited access to information, Resistance to change, Political interference, Inadequate monitoring and evaluation of policy outcome, Policy complexity means difficult for rural people to understand and implement and Insufficient stakeholder Involvement means lack of active involvement of local stakeholder in the policy development and implementation process.

To solve these concerns, a comprehensive plan incorporating targeted investments in human resources and infrastructure, sustainable resource management practices, and democratic decision-making is required. By acknowledging and addressing the specifics of the rural environment, policymakers and development professionals may support inclusive and resilient rural communities.

18.2 Unpacking the rural landscape reveals a spectrum of opportunities for sustainable development, community empowerment, and improved quality of life

18.2.1 Agro-ecotourism and Sustainable

Rural areas often boast natural beauty, cultural heritage, and unique landscapes that can attract tourists (Gursoy *et al.*, 2020). Encouraging agro-ecotourism and sustainable tourism projects can produce income, open up job opportunities, and protect regional customs and ecosystems. In rural areas, agro-ecotourism and sustainable tourism connect cultural diversity, agricultural vigour, and ecological sustainability in a way that works together. This strategy not only provides tourists with unique and genuine experiences but also strengthens the resilience and general well-being of rural communities.

18.2.2 Renewable Energy Development

Rural areas are rich in renewable energy resources such as solar, wind, and hydroelectric power (Makumbe *et al.*, 2018). Infrastructure for renewable energy is an asset that not only meets energy needs but also strengthens regional economies generates jobs and minimizes environmental effects. In order to advance sustainable development, improve energy access, and strengthen local communities, renewable energy development is essential in rural areas. In order to create a more sustainable, more resilient, and equitable energy future, rural communities may fully utilise renewable energy through strategic planning, investment, and involvement of stakeholders.

18.2.3 Value Added Agriculture and Agribusiness

Encouraging value-added agriculture practices and agribusiness ventures can enhance rural livelihoods and increase farm incomes (Khanal *et al.*, 2019). Agricultural products that are processed, packaged and sold locally or regionally boost rural economies and encourage entrepreneurship. Initiatives in agribusiness and value-added agriculture have a lot of potential to boost local food systems, sustainably create jobs, and boost rural economies. Rural communities may fully realise the promise of value-added agriculture and agribusiness to create strong, sustainable economies by utilising local resources, creativity, and teamwork.

18.2.4 ICT and Digital Connectivity

Improving access to information and communication technologies (ICT) and digital connectivity can bridge the rural-urban divide and facilitate economic and social development (Qiang *et al.*, 2019). Remote communities may have more chances thanks to ICT enabled services like telemedicine, online learning, and e-commerce. The opportunity for ICT and digital connectivity to empower rural communities, encourage inclusive development, and close the digital divide is enormous. ICT for rural areas can be modified by policymakers, development organisations, and private sector partners by addressing infrastructural issues, boosting digital literacy, lowering costs, and guaranteeing content relevance.

18.2.5 Community Based Natural Resource Management

Engaging local communities in natural resource management initiatives empowers them to conserve biodiversity, manage forests sustainably, and safeguard water resources (Ostrom, 2015). Community based approaches foster ownership, resilience, and equitable distribution of benefits. Community-based natural Resource Management is a holistic and inclusive approach that recognises the integral role of local communities in sustainable resource management. By fostering community empowerment, integrating traditional knowledge, and addressing challenges collaboratively, CBNRM can contribute to both conservation and the well-being of rural communities.

18.2.6 Skills Development and Capacity Building

Investing in skills development and capacity-building programs enhances human capital and empowers rural residents to seize economic opportunities (Wollmann & Kooy, 2019). Training in agricultural techniques, entrepreneurship, and vocational skills equips individuals for employment and self-reliance. Skills development and capacity building are integral components of rural development strategies aimed at unlocking human potential, fostering inclusive growth, and building resilient communities. By investing in education, training, and empowerment initiatives, governments, NGOs, and other stakeholders can unlock the latent talents and abilities of rural populations, driving sustainable development and prosperity.

18.2.7 Public Private Partnerships (PPPs) and Social Enterprises

Collaborative efforts between governments, private sector entities, and social enterprises can unlock synergies and mobilise resources for rural development (Klerkx *et al.*, 2019). PPPs and social entrepreneurs foster inclusive growth, scale up projects, and stimulate innovation. Together, social businesses and public-private partnerships (PPPs) may address development issues and encourage equitable growth in rural areas. PPPs and social entrepreneurs can help rural economies reach their full potential, enhance livelihoods, and create resilient communities by encouraging cooperation, innovation, and community involvement. Sustainable solutions that address social, economic, and environmental concerns in rural development agendas can be produced by integrating the two models.

Sustainable rural development can be established through the application of these opportunities in a participative and context-specific manner. Policies, nongovernmental organisations, and local communities can work together to improve the resilience and general well-being of rural areas by making use of these opportunities.

Table 18.1 Summary Table of Public-Private Partnerships (PPPs) and Social Enterprises

Factors	Challenges	Opportunities
Heterogeneity of Rural Areas	Variations in geography, culture, and economic activities. Inequalities in healthcare and education.	Modulation approach to address unique needs. Entrepreneurial ventures based on local resources.
Data Accessibility and Quality	Limited data due to remote locations and outdated methods.	Improved data collection methods.
Infrastructure Shortage	Inadequate roads, electricity, and telecommunications.	Investment in infrastructure for economic growth.
Economic Diversification	Dependence on agriculture. Low productivity. Vulnerability to climate change.	Value-added agriculture. Economic diversification.
Social Services and Healthcare Access	Limited access to quality healthcare and education.	Addressing disparities in service delivery.
Environmental Sustainability	Vulnerability to environmental degradation. Climate change impact.	Agro ecotourism and sustainable tourism. Renewable energy development.
Policy Implementation and Governance	Weak policy implementation. Lack of community participation.	Effective resource allocation.
Migration Trends	Youth migration to urban areas.	Demographic balance.
Skills Development and Capacity Building	Investment in human capital.	Vocational skills for employment.
Public Private Partnerships (PPPs) and Social Enterprises	Collaborative efforts are needed.	Inclusive growth and innovation.

18.3 Conclusion

The rural landscape, with all its complexities, is not merely a backdrop of challenges but a dynamic field of opportunities. Strategic and sensitive interventions that recognise and respect the unique characteristics of each region are crucial. By fostering a holistic approach that integrates technological advancements, educational initiatives, and sustainable practices, rural communities can navigate their way towards a more equitable and prosperous future.

References

Agrawal, A. & Gibson, C. C. (1999). "Enchantment and disenchantment: The role of community in natural resources conservation." World Development, 27(4): 629-649.

Berkes, F. (2004). "Rethinking community-based conservation." Conservation Biology, 18(3): 621-630.

Berkes, F. (2007). Community-Based Conservation in a Globalized World. Proceedings of the National Academy of Sciences, 104(39): 15188-15193.

Bhattacharyya, S. C. (2017). Rural Energy and Development. Springer.

Biswas, R. K., Kabir, E., & Khan, A. (2019). Community-based health Initiatives in Bangladesh: A Review of Interventions, Challenges, and Opportunities. Journal of Preventive Medicine and Public Health, 52(4): 211-219.

Bryden, J. (1994). Rural Development: Issues, Options, and Guidelines for Strategy. Oxford University Press.

Chen, R., & Huffman, W. (2019). Challenges and Opportunities in Rural Development: Perspectives from China. Journal of Rural Studies, 67: 107-115.

Cloke, P., & Edwards, G. (1986). Policy Failure and Administrative Decentralization in England and Wales. Regional Studies, 20(4): 309-322.

Cloke, P., & Milbourne, P. (1992). Deprivation and Class in Rural Areas: Some Implications of the 'New Times'. Journal of Rural Studies, 8(1): 35-50.

Dettmann, R. L., & Dimitri, C. (2007). "Local food system: concepts, impact, and issues." ERR97, U.S. Department of Agriculture, Economic Research Service.

Faggian, A., & McCann, P. (2009). Human capital and regional development. In Handbook of Regional Growth and Development Theories (pp. 403-426). Edward Elgar Publishing.

FAO. (2018). ValueAdded Agriculture: An Opportunity for Smallholder Farmers in Africa. Rome: Food and Agriculture Organization of the United Nations.

Gursoy, D., Chi, C. G. Q., & Lu, L. (2020). Impacts of Agroecotourism on Sustainable Rural Development: A Stakeholder Perspective. Journal of Sustainable Tourism, 28(2): 253-274.

Hemmerling, S., Krueger, T., & Linke, T. (2017). Challenges in Assessing Rural Water Services - A Literature Review. Water Research, 1(2): 59-70.

Honey, M. (2008). Ecotourism and Sustainable Development: Who Owns Paradise? (2nd ed.). Island Press.

Hossain, M. S., Rahman, M. S., & Islam, M. R. (2020). Challenges and Opportunities in Accessing Healthcare Services in Rural Areas: A Case Study of Bangladesh. Journal of Rural Health, 36(3): 389-398.

Key, N., Sadoulet, E., & de Janvry, A. (2000). "Transactions costs and agricultural household supply response." American Journal of Agricultural Economics, 82(2): 245-259.

Khanal, A. R., Mishra, A., & Khanal, A. R. (2019). Agribusiness for Sustainable Rural Development: Opportunities and Challenges. Journal of Agribusiness in Developing and Emerging Economies, 9(4): 458-476.

Klerkx, L., & Rose, D. (2019). Dealing with the Wicked Problem of Sustainable Intensification: The Role of PublicPrivate Partnerships in the AgriFood System. Current Opinion in Environmental Sustainability, 40: 35-40.

Kumar, S., & Singh, A. (2021). Adoption of Mobile Agriculture among Rural Farmers: Opportunities and Challenges. Telematics and Informatics, 57: 101-480.

Lobao, L., & Meyer, K. (2001). The Great Agricultural Transition: Crisis, Change, and Social Consequences of Twentieth Century US Farming. Annual Review of Sociology, 27: 103-124.

Mair, J., & Marti, I. (2006). "Social entrepreneurship research: Asource of explanation, prediction, and delight."Journal of World Business, 41(1): 36-44.

Makumbe, P., Mlambo, S., & Manase, G. (2018). Renewable Energy Technologies for Rural Development in SubSaharan Africa: A Review. Renewable and Sustainable Energy Reviews, 81(1): 1002-1016.

Martinez, S., Hand, M., Da Pra, M., Pollack, S., Ralston, K., Smith, T., & Newman, C. (2010)."Local food system: concepts, impact, and issues."ERR97, U.S. Department of Agriculture, Economic Research Service.

Matarrita Cascante, D. & Brennan, M.A. (2019). The Role of Entrepreneurship in Sustainable Rural Development: Perceptions from Rural Tourism Operators. Sustainability, 11(15): 40-93.

OECD. (2019). "PublicPrivate Partnership for Sustainable Development." Paris: OECD.

Ostrom, E. (2015). Governing the Commons: The Evolution of Institutions for Collective Action. Cambridge University Press.

Qiang, C. Z., Kuek, S. C., Dymond, A., & Esselaar, S. (2019). Universal Access to the Internet: Reality or Illusion? Information Technology for Development, 25(1): 521.

Rasul, G., & Sharma, B. R. (2016). Rural Infrastructure Development in India: Emerging Challenges and Opportunities. Journal of Rural Development, 35(4): 437-454.

Reardon, T., Barrett, C.B., Berdegue, J.A., & Swinnen, J.F. (2009). "Agrifood industry transformation and small farmers in developing countries." World development, 37(11): 1717-1727.

Reid, H., Alam, M., & Berger, R. (2019). Climate Change & Agriculture: Impacts and Adaptation. In Routledge Handbook of Agricultural Economics (pp. 188-207). Routledge.

Shahbaz, B., Arshad, H. M., & Shabbir, G. (2019). Challenges and Opportunities in Rural Development: A Case Study of Pakistan. International Journal of Rural Management, 15(1): 25-39.

United Nations Department of Economic and Social Affairs (2019). "World Economic Situation and Prospects 2019." New York: United Nations.

VanderZanden, A., Wirsing, A., & Arthur, S. (2019). Environmental Challenges and Opportunities in Rural Landscapes: A Global Perspective. Environmental Science & Policy, 94: 168-176.

Wollmann, C., & Kooy, M. (2019). Skilling India's Rural Women: Opportunities and Constraints. Journal of Rural Studies, 69: 47-57.

World Bank. (2007). World Development Report 2008: Agriculture for Development. World Bank Publications.

World Bank. (2020). World Development Report 2020: Trading for Development in the Age of Global Value Chains. Washington, DC: World Ba

Index